UX Research Methods for Media and Communication Studies

A comprehensive guide to qualitative research methods in user experience (UX), the interaction between humans and digital products, designed for media and communication students.

Angela M. Cirucci and Urszula M. Pruchniewska provide an accessible introduction to the field (including the history of UX and common UX design terminology). Readers are taken through the entire research design process, with an outline for preparing a study (including a planning template), a discussion of recruitment techniques, an exploration of ethics considerations, and a detailed breakdown of 12 essential UX research methods. The 12 methods covered include emotional journeys, screenshot diaries, walkthroughs, contextual inquiry, card sorting, and usability testing, with the chapter for each method including a step-by-step breakdown, discussions of in-person versus virtual procedures, and a "What You Need" section. Throughout the book, useful parallels are drawn between traditional academic research methods and UX methods, and special attention is paid to diversity and inclusivity.

This is an essential text for media and communications students wishing to become familiar with UX research methods, a rapidly growing field that will open numerous exciting career paths for graduates.

Angela M. Cirucci is Assistant Professor of Communication Studies at Rowan University. Her work explores the symbolic meaning of programming languages, the intersection of institutional practice and user knowledge, and user experience. She also examines how digital spaces impact the lives of marginalized communities. She frequently teaches classes on New Media and Communication, Participatory Media, User Experience Design and Research, and Communication Research Methods.

Urszula M. Pruchniewska is a Senior UX Researcher in the nonprofit industry. She is committed to promoting inclusivity and diversity in research and design practices and to using technology for positive social change. Previously, she was Assistant Professor of Communication Studies at Kutztown University, where she taught User Experience Design and Research, Communication Research Methods, Social Media Ethics, and Social Media Theory and Strategy.

UX Research Methods for Media and Communication Studies

An Introduction to Contemporary Qualitative Methods

Angela M. Cirucci and Urszula M. Pruchniewska

NEW YORK AND LONDON

Cover image: Getty images

First published 2022
by Routledge
605 Third Avenue, New York, NY 10158

and by Routledge
2 Park Square, Milton Park, Abingdon, Oxon, OX14 4RN

Routledge is an imprint of the Taylor & Francis Group, an informa business

© 2022 Angela M. Cirucci and Urszula M. Pruchniewska

Library of Congress Cataloging-in-Publication Data
A catalog record for this book has been requested

ISBN: 978-1-032-02078-5 (hbk)
ISBN: 978-1-032-01866-9 (pbk)
ISBN: 978-1-003-18175-0 (ebk)

DOI: 10.4324/9781003181750

Typeset in Times New Roman
by ApexCovantage, LLC

Access the Support Material: www.routledge.com/9781032018669

Contents

Figures

Section One

Introduction

1   What Is UX?

What is *user experience*? Think about something you love doing that requires a specific physical object, such as drinking coffee out of your favorite mug, running in your "fastest" sneakers, or writing with your special pen. What is this experience? Why do you love it? What is your favorite part of using this object? Now, think about a digital experience you love, such as using a particular website or app. What is it? Why do you love it? What is your favorite part of using this product? The answers to these questions form your *best* user experience. Of course, user experience can also be negative—think about a mug that is an awkward size for your hand or a website with buttons that are clickable but don't actually perform an action. These interactions are frustrating and unpleasant to the user. They might make you question who designed this and why they didn't think about how somebody would actually be using it? This is where the field of user experience comes in.

This book is all about user experience, or UX, which is paying attention to how a person (or "end-user" in technical terms) uses—that is, interacts with and experiences—a particular product or service. UX typically includes researchers and designers, though increasingly the field is becoming more complex, with additional roles such as those for UX writers and UX content strategists becoming more prevalent. A UX researcher is the person who conducts research to better understand the user and their needs before the product or service is created; they also conduct research to understand how a current product or service can be improved from the user perspective. A UX researcher is also the person who typically tests a designed product to make sure it actually meets user needs. A UX designer is the person who designs (and redesigns) the product based on the insights that the UX researcher provides. This book focuses on the UX of *digital products*, such as devices, websites, platforms, and apps, and specifically on the different ways of doing *research* to inform the creation of a satisfying user experience.

Maybe you have seen "UX" in a recent blog post or job ad. But, studying user experience has a surprisingly older history than you may think. It extends back much further than digital spaces and is rooted in buildings, cities, and tangible objects, slightly morphing, but keeping its foundational pieces, as it moves through early human-computer interaction (HCI) research and into

DOI: 10.4324/9781003181750-2

today's app-centric world. Large companies like Facebook and Google with hundreds of UX researchers, as well as smaller companies with only one UX researcher, all employ some of the methods in the upcoming chapters.

The goal of this book is twofold. While we want to help aspiring UX researchers learn the necessary skills for the job, we also see this book as a means of introducing new and exciting research methods into academia. These creative methods developed in UX can be used in Media and Communication Studies to better understand how and why different people interact with digital technologies (and to what ends) and how design impacts communication. Perhaps even more importantly, the methods outlined in this book provide a more applied perspective that parallels the design choices made for apps and websites, allowing academic researchers to ultimately conduct more relevant studies of digital products.

This book presents some of the most common qualitative UX research methods. These methods are creative, contemporary, and adaptable. But, as you will see, they are also reliant on traditional research skills like interviewing, observing, and focus group moderating. They also rely on the ability to think critically and outside the box, to find emergent trends, and to use previous findings to create informed, desirable, and feasible solutions to problems. In this introductory chapter, we introduce you to the field of UX research, tracing the history of the field and linking it to more traditional research, with which you may be a little more familiar.

Understanding User Experience

One of the best examples of UX comes via an unexpected product—ketchup. As Lisa Jewell explains, Heinz ketchup was introduced in 1876 and made very few changes to its classic glass bottle for over 100 years. Interested in learning how people were using ketchup at home, Heinz hired some researchers to visit families in their houses in the US and observe them at mealtimes. Unexpectedly, they realized that a design change was needed.[1]

Think about the *experience* of getting ketchup out of an old-school glass bottle. You have to get the cap off, turn the bottle upside down, and then shake or hit the bottle (on the bottom? on the "57?"), all the while remembering it is glass and you probably shouldn't drop it. Not surprisingly, Heinz's researchers found that serving ketchup was an adult task. This realization came via a few observations. First, the kids were not having a fun experience with the ketchup largely because they were not allowed to use the bottle. Second, because the parents were doling out the ketchup, very little ketchup was being used and this means that, of course, less ketchup was being purchased.[2]

The research output solutions spoke to both user experience and business viability. If the ketchup could be instead stored in a plastic container with which kids could be trusted, Heinz could provide a better user experience (while also increasing consumption, and so helping the bottom line). Heinz went on to design the EZ Squirt bottle, a plastic container with a special nozzle. It was lighter and the ketchup came out more cleanly. Families that purchased the new design increased ketchup consumption by 12 percent. Heinz didn't stop there. After more observations, they realized families would typically store their bottles upside down, so they flipped the label![3]

Here are three UX takeaways from this study, as outlined by Jewell. One: Heinz made changes based on *data* from users. Two: These changes solved an actual user experience problem. Three: Interview or survey data would have been much less likely to uncover the problem. Observing users in creative ways and authentic spaces trumps impersonal research, such as surveys.[4] Because this textbook is focused on empathetic and inclusive design, we want to add a fourth takeaway: The new design also made the bottle more accessible, as only one hand is now needed to serve ketchup.

The thing about user experience research is that you have to research *users' experiences*. It's right there in the name. Consumers use products in specific contexts and particular ways, and they use products differently, depending on who they are and what they need. No user experience is the same, but the same factors contribute to good user experience across use cases and contexts—we outline these in the next section.

What Makes Up User Experience

User experience includes both the look and feel of a product, as well as the usability—how easy and functional it is to use. Not only must a product be aesthetically pleasing and usable, it must meet the user's needs. The most beautiful product that works smoothly but doesn't ultimately serve a purpose is bad UX. It's also important to remember that any product must also serve the business needs or objectives of the company that creates it, not just the user needs.

According to Peter Morville, a pioneer in the UX industry, there are seven distinct factors that make up a good user experience (presented in Figure 1.1 as a honeycomb diagram that represents the "sweet spot" of UX):

1. Useful—Is the product useful? Does it have a purpose? Note that what is considered "useful" changes person by person, but generally a product should have a clear use for some defined target audience.
2. Usable—Can users effectively and efficiently achieve their end objective with the product?
3. Findable—Is content in the product or are the features of the product easy to find? Websites, for example, should be easily navigable by novice users.
4. Credible—Does the design enhance credibility of the product, that is, do the users trust the product?
5. Accessible—Does the product provide an experience that can be accessed by users with a full range of abilities?
6. Desirable—Is the product design and experience aesthetically pleasing? This facet includes things like branding, image, and identity.

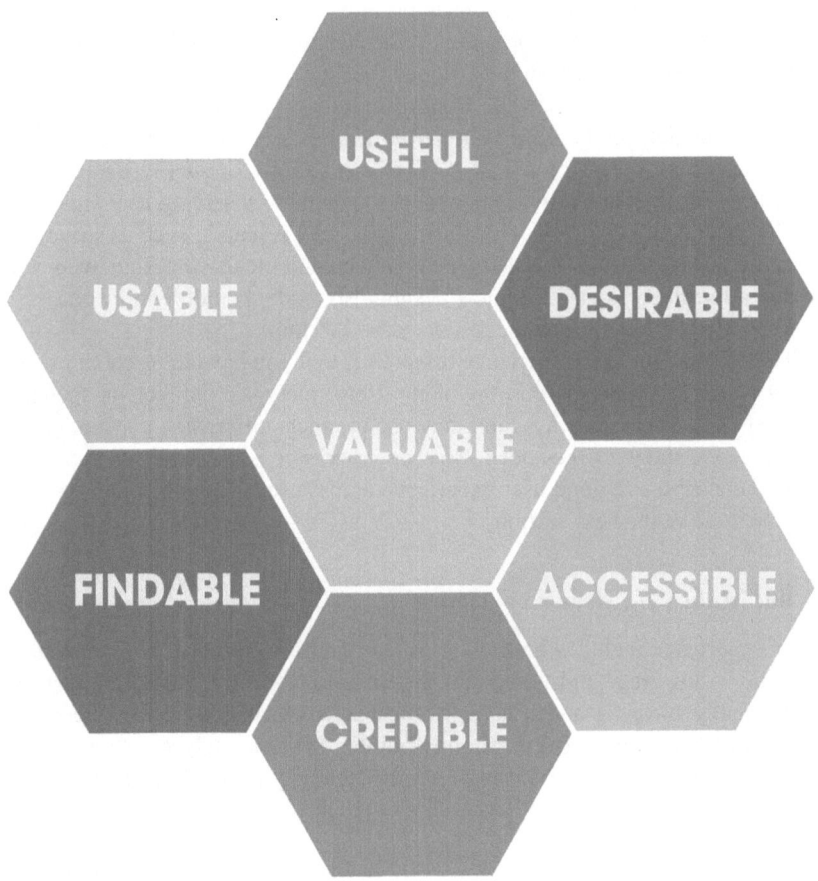

Figure 1.1 Morville's honeycomb of the seven distinct factors that make for good UX.

7.   Valuable—Does the product deliver value to the user *and* to the business? The product should deliver customer satisfaction but should also contribute to the bottom line (in for-profit companies) or to the mission and values (in nonprofits).[5]

The goal of UX is to meet the needs of the end-user through simple and elegant design that makes the product feel like a joy to use and to own. It's important to note that UX includes *all* aspects of an end-user's interaction with the company, its services, and its products—including marketing of the product and technical assistance or customer support when something goes wrong with the product. As Don Norman, the inventor of the term "user experience," explains:

No product is an island. A product is more than the product. It is a cohesive, integrated set of experiences. Think through all of the stages of a product or service—from initial intentions through final reflections, from first usage to help, service, and maintenance. Make them all work together seamlessly.[6]

This complete experience clearly goes way beyond just the UI (user interface), the specific point at which the user interacts with a product, such as a web page. Obviously, the interface is very important in a user's experience, but even if an interface is nice to look at, it could still contain flaws and inconsistencies that ruin the overall product or brand experience.[7] We get more into UIs in Chapter 4.

User Experience Versus Usability

Sometimes user experience and usability are used interchangeably, so it's worth exploring the differences in more depth here. Usability is one factor of user experience (as described by Morville above) that specifically targets *ease of use*. The ISO (the International Organization for Standardization) defines usability as:

> The extent to which a product can be used by specified users to achieve specified goals with effectiveness, efficiency, and satisfaction in a specified context of use.[8]

In other words, usability is focused on making a space that is easy for the user to become familiar with and competent in, as well as making it easy to reach their objective. Usability is certainly an important piece of UX, but it isn't exactly the same thing. User experience includes usability but is also so much more—including the experience beyond the product itself (such as customer service) and aesthetics (how pleasant the product is to look at).

The Digital Product Life Cycle and Research

Creating a good user experience in a digital product, such as a website, or app, involves a process that starts with research. Research means collecting and analyzing information to better understand something. In the case of UX research, it's about gathering information to understand user needs, desires, and pain points. In industry, this early research is often called discovery research or generative research, in that it generates a deeper understanding of the user and the research problem.

Once there are some research findings produced about what users need and want, the product (or new features for an existing product) can start being designed. Design includes thinking about how the website or app will look and how a user will navigate through it. During this stage, UX designers will create mock-ups, wireframes (low fidelity, simple versions of the final product), and prototypes (high fidelity, more detailed versions of the final product) that are tested with users to gather feedback on the design concept and make changes as necessary. This iterative testing is a type of research activity, as it involves collecting and analyzing information to ensure that the envisioned user experience is on track.

Next, the product is developed—in the case of digital products, this means adding back-end coding and functionality to a website or app. Finally, once the product is developed and launched (put out into the world for users to use), it is again tested, to make sure it effectively meets user needs. This stage of research is often called evaluative research, because it evaluates or tests whether the product actually does what it is supposed to. This UX process for digital products roughly follows the Design Thinking Mindset, a process for creative problem solving, explained in depth in Chapter 5.

Diving Further Into UX Research

Research is everywhere. We can consider taking the time to Google an idea and finding the most relevant and reliable results "research." Sure, it is perhaps less formal and maybe less rigorous, but, at its heart, it is still research. Maybe you have learned about lab research, employing the scientific method and testing hypotheses. Or, maybe when you think about research you think about long and difficult-to-read papers found in academic journals through Google Scholar or EBSCO. It's important to remember that research is broader—and some type of research is a part of almost every job. So, even though this book is geared toward aspiring UX researchers and Media and Communication Studies students, you should be aware of the wider applications of research. We are *always* doing some form of research, but here we will focus on how to make that research credible, valid, and inclusive, and how to choose the best processes that lead to desirable, feasible, and viable changes to users' experiences.

Typically, especially in academia, research is broken into two prongs—quantitative and qualitative. Quantitative research in Media and Communication Studies includes closed surveys and experiments, with researchers interested in measuring behaviors and attitudes or testing relationships between variables. Qualitative research instead uses interviews, open-ended surveys, focus groups, and observations. Qualitative research is often more exploratory (when the researcher doesn't fully know yet what exactly they want to research, they just want to see what's out there that could be interesting to research in more depth), and qualitative methods are used when the researcher wants to deeply understand some group of people or culture. This book is about *qualitative* research methods common in UX, such as contextual inquiry, card sorting, and usability testing.

You may have heard of other members of design teams like graphic designers, information architects, and interaction designers. A UX researcher is typically a part of the design team, and while a UX researcher is not expected to be a designer or a programmer, they are expected to understand the expectations and limits of everyone else on the team (this is why we have a chapter on the design elements of interfaces, Chapter 4). Being a great UX researcher means understanding how to conduct great research, while at the same time following the company's business model and achieving organizational goals.

This Book

In this book, we will mostly focus primarily on the UX of popular websites and apps. But, user experience research originates from studying how people experience physical spaces and experiences. With that in mind, the methods in this book can be applied to a myriad of digital products, including games, operating systems, computer programs, and hardware or devices. Particularly relevant to fields like Media and Communication Studies, the UX research we cover in this book is mostly concerned with better understanding how apps and websites communicate with end-users.

Again, our main goal for this book is twofold. First, we want to help Media and Communication Studies students prepare for the workforce. UX researcher jobs are abundant, ranging from positions at top digital corporations like Facebook and Google, to smaller companies looking for one or two researchers, and to consulting companies that exist solely to conduct UX studies for a variety of clients.

Second, we are passionate about updating and energizing qualitative research methods within the Media and Communication Studies fields. Many research methods textbooks think of digital spaces primarily as recruitment tools—you can use a Facebook group or a tweet to find participants. Others think of them as places to scrape data (information about and from users), often not taking into consideration privacy issues and merely assuming people's words and images are up for grabs because they are posted "publicly." We want to promote innovative techniques for gathering and analyzing data not just *from* but also *about* digital spaces and products—and the people who use them! So, while the book is focused on UX in the sense that we are introducing techniques popular in industry research, the majority of the methods outlined in this book can also be used creatively for academic research, research that adds to a body of knowledge about digital technologies and their users.

Even though there are many exciting, innovative methods in the UX field, a good UX researcher still relies heavily on foundational qualitative research skills that are typically used in academic research. As you will see in the upcoming chapters, observation, asking good questions, and thinking critically all are essential to conducting great UX research. Almost every method we outline in this book incorporates some type of interviewing or observing, which is why we dedicate a chapter (Chapter 8) to a brief overview of the most common traditional qualitative methods: interviews, observations, open-ended surveys, and focus groups.

Overview

In the following chapters that comprise Section One, we introduce you to the world of digital structures and the study of how humans interact with them.

In Chapter 2, we cover the very important and timely topic of bias in digital structures and UX. We discuss the ways in which norms and goals are "baked in" to human-created spaces, including how objects can have implications for users online, as well as in their everyday "offline" lives. We cover how old and new technologies have affected those who identify with marginalized communities.

In Chapter 3, we briefly trace early digital design that was less about interactivity and more about qualities like usefulness, desirableness, and affordability. Early computers were programmable machines that only viewed people as one component of the production system. Command line interfaces, early mice and keyboards, VISICALC, and WordStar are explored here. In Chapter 3 we also provide more contemporary definitions of design that place users in the center and that situate UX researchers' goals in striving to create experiences that fit diverse user contexts, characteristics, and patterns of life. Chapter 4 covers important components related to UX research—interfaces, navigational components, and other interface elements such as tooltips and coach marks. While interface design fundamentals are more prominent in UX *design* training, UX researchers are expected to be familiar with these concepts.

Section Two is devoted to helping you understand how to best plan a UX study. In Chapter 5 we introduce the guiding process for this book—the Design Thinking Mindset. This process involves five main stages of Design Thinking—empathizing with users, defining problems, coming up with ideas, designing/building prototypes, and testing those prototypes. Chapter 6 then provides a walkthrough of the process of actually planning a UX research project. We discuss gathering background information, logistics, recruitment techniques, and ethical and accessibility considerations. Chapter 7 covers reporting on and presenting your research findings. We compare academic and industry reports and also outline how to present UX findings well, discussing presentation tools such as Microsoft PowerPoint, Canva, Mural, and Zoom.

In Section Three we finally get to the exciting UX research methods! We cover 12 methods, representing all five stages of the Design Thinking Mindset. But, we begin the section with an overview of foundational qualitative research methods.

- Chapter 8—this chapter provides a refresher of traditional qualitative research methods, including interviews, focus groups, qualitative surveys, and observations. These are crucial skills to have for *any* research activities, and the UX methods in this book rely on a mastery of these foundational skills.
- Chapter 9—Emotional Journey Mapping—this method is about using touchpoints of a process within an app or a website to map users' emotional experiences on a linear axis. Emotional journey maps reveal how participants *feel* while using your product.
- Chapter 10—Screenshot Diaries—this is similar to the traditional method of diary studies (collecting information about participants through them keeping diaries about their daily activities), but users are asked to

document certain activities and events or salient topics through screen-shots and captions. Screenshot diaries allow researchers to observe how a participant lives their lives alongside your digital product, highlighting important features and functionalities.

- Chapter 11—Breakup and Love Letters—this method is about users writing creative letters to products describing why a user stays (and continues using the product) or why they may leave. Breakup and love letters elicit feelings based on real-life experiences and salient moments with particular digital products.
- Chapter 12—Contextual Inquiry—here, the researcher combines observations of a user interacting with a product in a natural environment (how they would typically use the product in real life) with interviews, asking the user why they do specific things with the product. Contextual inquiry is great for providing insights into both how people think about their interactions with digital devices and what they actually do during these interactions.
- Chapter 13—Personas and Scenarios—these are fictional but realistic representations of a product's end-users and the narratives around their use of a product. Personas and scenarios allow researchers to synthesize information to more easily understand user habits and preferences.
- Chapter 14—Problem Trees—this mapping technique for the Define stage of Design Thinking helps researchers to understand an issue by analyzing its causes and effects. Problem trees help to create a research problem and guide the remainder of the research project.
- Chapter 15—Cognitive Mapping—here, users are asked to create mental maps of an app or website by sketching, drawing, graphing, or charting. Specifically, cognitive mapping can be used in the Ideate phase to illustrate what users enjoy and would like to see changed within the product.
- Chapter 16—Brain Writing and Darkside Writing—these methods are similar to focus groups but are used in the Ideate phase. Brain writing is like brainstorming and asks participants to write down ideas, pass around the ideas, and add on to or improve as the paper moves around the room. Darkside writing flips the Problem Statement into its opposite and asks participants to propose a "solution." These "solutions" are then flipped back to provide new and creative solutions to the problem.
- Chapter 17—Card Sorting and Tree Testing—these methods visualize potential structures and organization of content in a digital product. Card sorting involves giving participants cards with content on them and then watching as they group or order these cards into a meaningful structure. Tree testing is the opposite—participants are asked to navigate through a provided structure (or tree) and to reveal findability of specific content.
- Chapter 18—Heuristic Evaluations and Walkthroughs as Critical Analysis—here, researchers (or other experts) systematically navigate through the website or app with the goal of evaluating it according to predetermined

criteria. These methods are used for finding usability issues or other problem areas in a particular digital product.

- Chapter 19—Usability Testing—in usability testing (unsurprisingly done during the Testing phase), researchers test prototypes of a digital product with real users, to see whether the finished website or app is user-friendly and easily navigable.
- Chapter 20—A/B Testing—this is comparing two versions of a website or an app with users. A/B testing allows researchers to ask participants their preference between a prototype and a current version of an app or a website or two new proposed designs.

Each UX method chapter includes a step-by-step breakdown as well as a "What You Need" section that outlines tools and resources considered generally necessary to use that method. You will find that, for most methods, this section is relatively short. Surprisingly, even though we are researching experiences related to complex apps and websites, the studies themselves do not require many tools. We try to suggest a myriad of tools throughout the book, and most are free or available through most colleges and universities. While there are more advanced, and expensive, tools that UX researchers may use, we have decided to not suggest any programs that require pricy subscriptions or steep learning curves.

We also include a case study in every chapter. These case studies provide a real-world example of that particular method being applied in either academia or industry. Note that there are many more examples for you to look at on your own! You can find more examples of research studies using different UX methods here:

- NNgroup.com
- UXplanet.org
- Interaction-design.org
- boxesandarrows.com
- Journal of Usability Studies
- Journal of Human-Computer Interaction
- UX Research Blogs

We also briefly discuss how to practically use each particular method in-person *and* virtually, given the reality of our even-more-digital world since the COVID-19 pandemic. We end with discussion questions to get your brain juices flowing. Now, let's get your UX research journey started in Chapter 2, which focuses on the critical consideration of bias in UX research and design.

Notes

1. Lisa Jewell, "User Research-What's Tomato Ketchup Got to Do with It?" *UX Planet*, May 14, 2018, https://uxplanet.org/user-research-whats-tomato-ketchup-got-to-do-with-it-758bfb536ca3.
2. Ibid.

3. Ibid.

4. Ibid.

5. Peter Morville, "User Experience Design," *Semantic Studios*, June 21, 2004, https://semanticstudios.com/user_experience_design/.

6. Donald A. Norman, "The Way I See It: Systems Thinking: A Product Is More Than the Product," *Interactions* 16, no. 5 (2009): 54.

7. Don Norman and Jakob Nielsen, "The Definition of User Experience (UX)," *Nielsen Norman Group*, accessed July 18, 2021, www.nngroup.com/articles/definition-user-experience/.

8. "ISO 9241–11:2018(en)," *ISO*, 2018, www.iso.org/obp/ui/#iso:std:iso:9241:-11:ed-2:v1:en.

2   Biased Digital Structures and UX

This book is based on a core assumption: digital products are *not* neutral. Social media, such as Facebook, Instagram, and Snapchat, are not "just neutral platforms" for people to communicate through, nor are *any* digital products, including devices (such as iPhones), websites, or apps. Things that are designed and made by people can never be neutral, as people themselves are not neutral, and the society we live in is certainly not neutral. Designers, researchers, and engineers all have their own beliefs and values, and live in environments steeped in cultural and societal norms and assumptions. These biases, values, and assumptions seep into the digital products they work on, whether consciously or unconsciously. In this chapter, we talk about politics, power imbalances, and bias and how these concepts feature in UX research and design.

Usually we discuss people having politics. But, in his popular piece from 1980, "Do Artifacts Have Politics?," Langdon Winner argues that technical things have political qualities. Specifically, Winner discusses machines, structures, and systems that embody forms of power and authority. He talks about how technologies are shaped by social and economic forces, and how the values and politics of these forces reside *within* the technologies themselves. Even though human-made objects do not have agency, people are biased, and it is inevitable that they bake in these biases when they create or design products. So, it's important to pay attention to the characteristics of technical objects and the meaning of those characteristics—*why* are things made the way they are? What assumptions do things embody? And what outcomes do things create through their use?[1]

For example, Winner explains, consider the overpasses in Long Island. These bridges over the parkways of New York are very low, with some having as little as nine feet of height clearance. Even though the constrained height of these bridges might not be noticeable at first glance, these overpasses are low enough that buses can't drive under them. These 200 or so overpasses were intentionally designed to be low by Robert Moses, the master builder and architect behind much of New York's transport infrastructure from the 1920s to the 1970s. Moses wanted to discourage buses on his parkways, so that only people who owned cars (i.e., white, middle/upper-class people) could use them

DOI: 10.4324/9781003181750-3

freely for leisure and recreation. Anyone who used public transit—primarily poor people and people of color—did not have easy access to the parkways. As you can see, Moses's design decisions had clear discriminatory outcomes, and this bias lives on in the things—the parkways themselves—long after Moses died, making them inherently not neutral.[2]

But surely most designers, builders, architects, and engineers aren't deliberately prejudiced like Moses? Unfortunately, as Winner points out, bias is often baked into objects without intention. He highlights how much of the human-made physical world is (and was even more so in the 20th century) inaccessible to people with disabilities. This is less about intentional exclusion and more about neglect stemming from the limited experience of able-bodied architects and builders creating a world based on their own needs and experiences.[3]

Some technologies aren't inherently exclusionary to particular groups of people but have unintended real-world consequences that can be harmful. Winner provides the example of the mechanical tomato harvester created in the 1940s. This machine was much more efficient at picking tomatoes than doing it by hand, but it was also much rougher on the plants, by shaking the tomatoes loose from their stalks. This led to the need to develop new tomato varieties that were able to withstand the motion. These new breeds, coupled with the high cost of the equipment, fundamentally changed how tomatoes are grown, displacing smaller farms and contributing to the rise of large agri-business.[4]

Technologies build order into our world; they structure human activity. Societies choose structures for technologies that influence how people are going to work, communicate, travel, and consume. This happens over long periods of time. But it's important to clarify here that our assumptions in this book are not based on technological determinism, the idea that technology shapes the development of society, forcing social change by its very design. Rather, we recognize that yes, technologies influence and shape society but they themselves are also shaped by social and economic forces.

Ultimately, we want to highlight that *people* are behind technologies. They make certain choices in the research and design process, and these choices are based in the socio-cultural context, their goals, and, often, who is paying for the design and development of a product. Frequently, these choices aren't even conscious. Different people are differently situated and possess unequal degrees of power and unequal levels of awareness—and these differences are embedded in the products they create.

Many scholars and industry experts have more recently touched on this idea of digital technologies having prejudices and biases baked in—and have highlighted the need for more inclusivity, diversity, and compassion in technology research, design, and use. In this chapter, we briefly summarize a few scholars who have contributed important work to the critical analysis of technologies. We also include industry discussions on how to make applied research and design less biased and more accessible to all.

Judy Wajcman and Gender Biases

Judy Wajcman specifically focuses on gender and the ways in which technologies reflect gender inequalities. Not only do men still have a monopoly in the tech industry but gender inequalities are embedded in the technologies.[5] Drawing from Judith Butler,[6] Wajcman argues that social relations are "materialised in tools and techniques."[7] The tools that users fold into their lives are socially constructed, based in prejudices that already exist, privileging those who make and fund them (mostly straight, able-bodied white men).

For example, Bumble is a dating app designed for women to take control of the dating game, by only allowing the woman to start the conversation after matching with a potential partner. Scholars Rena Bivens and Anna Shah Hoque[8] and Caitlin MacLeod and Victoria McArthur[9] studied the interface of Bumble and found that the app makes normative assumptions about gender, sex, and sexuality. For example, at the time of their studies, the gender choices when creating a profile on Bumble were "male" or "female," and users could choose whether they were interested in "males," "females," or "both." (Bumble has since changed these options to include non-binary gender identification.)

In addition, the entire premise of Bumble is that men and women are most often trying to date each other. Thus, Bumble operates on a "heterosexual matrix,"[10] the idea that bodies identified as female at birth perform feminine gender identity and are attracted to bodies designated as male at birth that perform masculinity. Static binary logics (male/female, heterosexual/homosexual) permeate Bumble's design, and, as such, the app is "optimized for straight cisgender women."[11] Who would have thought that the "feminist" dating app could still be so exclusionary by its very design to a large swathe of people!

Safiya Noble and Racist Structures

Safiya Noble's 2018 book, *Algorithms of Oppression*, presents a meticulous analysis of Google's search engine, providing evidence for the ways in which the algorithm is racist. She begins the book by writing about one particularly shocking example, an example that happened to inspire the research for her book. In fall 2010 she was searching the web looking for something that may interest her stepdaughters and nieces. She typed "Black girls" in Google's search bar, and the first result was porn.[12]

In 2012, Google updated their algorithm and "Black girls" no longer yielded immediate porn links, but typing searches like "Latinas" and "Asian girls" still did. Noble includes other examples to back up her claims, including a Google image search for "gorillas" that suggested pictures of Black people, a related search suggestion for "Michelle Obama ape" when searching "Michelle Obama," and a 2016 Google maps search for "[extremely racist Black slur] house" that pulled up the White House. (Remember that during this time Barack Obama was president of the US.)[13]

As Noble explains, it is very easy for Google, and other companies, to say that instances like these are just "glitches" or "bugs." And, once they are made public, Google works quickly to fix the "algorithm's mistake." But, there is a clear lack of respect and regard for women and people of color when these types of results are even possible. It is also clear that UX research and testing did not consider these instances; otherwise, they would never have been possible in the first place. It doesn't help, as Noble includes, that Google has a history of employing engineers that openly promote racist and sexist attitudes.[14]

Sara Wachter-Boettcher and Toxic Tech

In her book *Technically Wrong*, Sara Wachter-Boettcher shows how the values, biases, and experiences of people who design technologies make their way into a myriad of tech products, including social media and search algorithms. For instance, a common feature of social media is retrospective "memory" posts, such as "Year in Review," where Facebook automatically shows users their past year's content, such as their most-interacted-with posts. This is a design feature that is meant to remind users about the highlights of their year and bring up the associated happy feelings.[15]

However, as Wachter-Boettcher prompts readers to consider, what if a user's most "interacted with" post was a sad event that their friends "liked" and commented on to show support? This was the case with Wachter-Boettcher's colleague, who got this unwelcome reminder of his young daughter's death from cancer in such a "year in review." The designers of the feature clearly did not consider that different people might have different experiences—and that it's important to consider a multitude of scenarios when designing a feature that is meant to be widely used by a range of people. It's also important to allow such features to be customizable, so that people can opt out of possibly harmful or triggering features in an app.[16]

Wachter-Boettcher argues that representation matters—it's really important for a diverse group of people to be on teams designing products, so that multiple viewpoints and experiences are considered when creating tech. However, it's not just the people who work at tech companies that create "toxic tech" but tech culture itself—a culture that rewards doing things fast and being confident in your choices (often with no thought for wider ramifications). A related problem is the idea that the data that drives tech (such as analytics that inform algorithms and data that is fed into Artificial Intelligence [AI]) is objective and neutral—when it is not; machines learn from and are programmed by humans who have intrinsic biases and who are steeped in non-neutral cultural norms and expectations.[17]

Cathy O'Neil and Weapons of Math Destruction

Inherent biases are, of course, not just found in social media apps but all digital products. In her book *Weapons of Math Destruction*, Cathy O'Neil explains that digital products are seen as objective and neutral because they are based in math and science. This makes digital products difficult to question even though they are so deeply embedded in users' lives. And the largely opaque processes carried out by digital companies don't make it any easier for users to understand all that goes into creating the content they consume daily.[18]

O'Neil argues that high-powered tech companies are tautological—they get to define their own realities and then use these "realities" to justify their decisions and results.[19] When programming languages create digital spaces, the goal is to make models that represent offline objects. No model can be perfect because the digitization process necessarily simplifies the richness that people and things possess offline. So, social media apps, for instance, miss the complexity of people and the nuance of human communication.[20] But, importantly, these blind spots are not random; they "reflect the judgements and priorities of [their] creators," and they are "opinions embedded in mathematics."[21]

Sticking with the social media example, companies like Facebook profit from users' data because they then can use that data to rank, categorize, and score users, based on hundreds of models. These algorithmic processes "reveal" preferences and patterns that are then used to prey on users, attempting to get them to fall for some scheme, tapping into their insecurities, desperations, or ignorance.[22] Again, the data spawn new data that just work to justify the same processes, creating an infinite positive feedback loop. These patterns are not just used to justify targeted ads and social suggestions. They are used to decide things like who should go to jail, who should get a loan, FICO scores, and who should get what type of insurance.[23] Clearly, what happens online has consequences for people's offline lives.

Mar Hicks and a History of Domination

Bias in tech is not a new phenomenon. Looking back into computing history, we learn that computers have always been situated in domination. Mar Hicks explains, in their essay "When Did the Fire Start?," that early research into computers began with warfare. Computers were a US military project. For example, during World War II Colossus was built to break Nazi coded messages. But, women, specifically Black women, were doing the "dirty work." Women operated computers, but their important contributions were largely hidden and silenced. It is only recently that their stories have begun to emerge.[24]

At the time, there was an optimism around computers, an idea that computing experts could tackle societal problems, relying on governmental funding. This ethos is arguably still strong today. But computers always have privileged, and still privilege, those with the most money and power. The technologies that allow computers and the internet to function continue to be built on existing, problematic, power structures. Just because there is technological progress does not mean there is social progress. Accountability is sorely needed.[25] Corporations, like Facebook and Google, get to act like governments, not only in their wealth but in their power and reach.[26]

Adrienne Shaw and Finding Room for Resistance

Adrienne Shaw employs the work of Stuart Hall, a famous cultural studies scholar, to explain how people can interact with different digital technologies based on the power they hold. Hall argued that audiences are not passive dupes who simply absorb media messages delivered to them—they can interpret messages (or parts of messages) differently than the producer of the message intended, or even reject messages completely.[27]

Drawing on Hall, Shaw suggests the term "using positions" when thinking about technology use: dominant use (technology is used as intended by designers), negotiated use (used correctly but not exactly as intended), and oppositional use (unexpected use of technology). In other words, technologies shape and are shaped by social interaction, and users can do various things with and to technology, within boundaries (for instance, the constraints of technological features—what is actually physically possible).[28]

As Shaw points out, "what counts as a dominant, negotiated, or oppositional use is intrinsically linked to who has the power to define how technologies should be used."[29] Creators of apps define how the technology should be used, through design and functionality choices (such as the placement of buttons in the interface), as well as through marketing and branding, creating an expected culture of use. However, users can and often do interact with the technology in unexpected ways.

For instance, Gaby David and Carolina Cambre showed how Tinder users use the app's features in creative ways to bypass its limitations. For instance, Tinder allows limited picture uploads in the user profile. To overcome this constraint, users provide links to other social media profiles (e.g., their Instagram) in their Tinder profiles to showcase more pictures. How users decode spaces is largely based on their environment—who they are, what they are experienced, and what they believe. As we mentioned at the beginning of the chapter, this is not to be confused with technological determinism. Designs do not tell people what to think or do, but they do shape, in part, *how* people think and what people *can* do given the material and cultural constraints of technologies.[30]

UX Takeaways

So, what does this all mean for the subject of this book: UX research? Hopefully you can see that one of the key ways of fighting bias and prejudice in the design of digital products and spaces is to simply *do research* in the first place! This is why UX research is crucial in industry—to make sure that tech products are inclusive, equitable, and sensitive to a variety of scenarios and users. Even though we as researchers are not directly designing the website or app, we are undoubtedly shaping how digital spaces are structured—what is possible, what things look like, who is included, etc.

Actively researching—talking with, observing, empathizing with—the people who actually use digital products is a great way of making sure that you're not just designing for yourself, based on your own assumptions, ideas, and values. It's important to do research before you start creating a website or an app, but it's just as important to continue doing research during the design process and even after the product is launched. You want to make sure you are considering your users' needs at all times.

Of course, in order to be truly empathetic and diverse, you need to consult a diverse user base—that is, talk to lots of different people who might end up using your product! There's no point in only recruiting users for research who look and think the same way you do—they will only validate your biased assumptions. Remember that different people also have different ideas and experiences from each other—a straight, Black, upper-class, middle-aged woman living in a large city will probably have different experiences, viewpoints, needs, and goals for using a particular digital product than a low-income, Latina grandmother of six living in a rural area. It's also helpful to have a diverse team of researchers and designers, so that different viewpoints and experiences are already at the table internally throughout the product development process.

We also need to consider our own biases as researchers, beyond making sure we do user research with a diverse group of people. Consider:

- Are we looking for specific answers when analyzing our research and not letting the data guide us?
- Are we being as objective as possible and truly putting our users first (and not our own feelings) when thinking about design solutions?
- How can we make sure we are methodical and that our results are valid? If possible, it is helpful to have multiple researchers working on the same project, to validate insights gathered from findings. If it's not possible to have multiple people conducting the research (i.e., gathering data through interviews, observations, or other methods outlined in this book), it's important that multiple people can analyze the data and come up with their own conclusions—and then consolidate them as a team.
- We need to make sure to use the right methods for the questions we're asking and that we are nuanced in our analysis (e.g., balancing what an interviewee tells us versus what they show us during observations in the conclusions that we draw).
- If possible, we should validate our insights with the people we gathered data from. Go back to your research participants and ask them if the conclusions that you drew from the study resonates with them. Why or why not?

In addition to collecting and analyzing lots of data from current and potential users themselves, UX teams can also be deliberate in creating their designs. Make sure to:

- Pre-empt a variety of possible issues that might arise once people start using your product and work possible solutions into your design (this is called proactive rather than reactive design)

• Brainstorm lots of different scenarios or contexts of use with your team (think of the Facebook "Year in Review" example from earlier in this chapter)

  – Here, it's important to consider what assumptions we are making about how people will use our product. What happens if these assumptions are wrong?

• Have an awareness of history, culture, and social issues related to your product. No website, app, or device exists in a vacuum or is used by people with no history

Overall, in UX research and design, it's important to think of our users as whole human beings, with experiences, viewpoints, and lives outside of interacting with our product that we should ethically be considering. We must constantly remind ourselves that *users are people* to ensure our research and designs are not only useful but also inclusive, equitable, and ethical. Check out the resources given here for some more guidance on how to conduct ethical qualitative UX research:

The Code for America Qualitative Research Practice Guide: https://f.hubs potusercontent30.net/hubfs/5622333/CFA_QualitativeResearchGuide_v1.pdf

Chicago Beyond "Why am I always being researched" Guide (Equity Series Vol 1): https://chicagobeyond.org/researchequity/

Notes

1. Langdon Winner, "Do Artifacts Have Politics?" *Daedalus* 109, no. 1 (1980), www.jstor.org/stable/20024652.
2. Ibid.
3. Ibid.
4. Ibid.
5. Judy Wajcman, "Feminist Theories of Technology," *Cambridge Journal of Economics* 34, no. 1 (2010): 146, https://doi.org/10.1093/cje/ben057.
6. Read more on the concept of "performativity" here: Judith Butler, *Gender Trouble* (New York: Routledge, 1999).
7. Wajcman, "Feminist Theories of Technology," 147.
8. Rena Bivens and Anna Shah Hoque, "Programming Sex, Gender, and Sexuality: Infrastructural Failures in the 'Feminist' Dating App Bumble," *Canadian Journal of Communication* 43, no. 3 (2018), doi:10.22230/cjc.2019v44n3a3375.
9. Caitlin McLeod and Victoria McArthur, "The Construction of Gender in Dating Apps: An Interface Analysis of Tinder and Bumble," *Feminist Media Studies* 19, no. 6 (2019), https://doi.org/10.1080/14680777.2018.1494618.
10. Bivens and Hoque, "Programming Sex, Gender, and Sexuality," 449.
11. Ibid.
12. Safiya Umoja Noble, *Algorithms of Oppression* (New York: NYU Press, 2018).
13. Ibid.
14. Ibid.
15. Sara Wachter-Boettcher, *Technically Wrong* (New York: WW Norton & Company, 2017).
16. Ibid.
17. Ibid.

18. Cathy O'Neil, *Weapons of Math Destruction* (New York: Crown Publishing Group, 2016).
19. Ibid., 7.
20. Ibid., 20.
21. Ibid., 21.
22. Ibid., 70.
23. Ibid.
24. Mar Hicks, "When Did the Fire Start?" in *Your Computer Is On Fire*, eds. Thomas S. Mullaney, Benjamin Peters, Mar Hicks, and Kavita Philip (Cambridge: MIT Press, 2021), 11–25.
25. Ibid., 20.
26. Ibid.
27. Adrienne Shaw, "Encoding and Decoding Affordances: "Stuart Hall and Interactive Media Technologies," *Media, Culture, & Society* 39, no. 4 (2017).
28. Ibid.
29. Ibid., 8.
30. Gaby David and Carolina Cambre, "Screened Intimacies: Tinder and the Swipe Logic," *Social Media + Society* 2, no. 2 (2016), https://doi.org/10.1177/20563051 16641976.

3 How Did We Get Here?

With each new generation, whatever is prevalent in the digital world at the time becomes "normal" to that group. If you grew up with the internet, for instance, having the internet feels normal. It is an expected utility in your life. Think about younger generations; what will "normal" be for them? Now think about your parents and grandparents. You can see why there is often conflict; their "normal," or what the popular technologies were when they were growing up, is quite different from your "normal."

These considerations are also important from a diversity and inclusivity perspective. With each new generation, the things we continue to allow to be "normal"—sexism, racism, ableism, etc.—will continue to feel acceptable. However, if we work to make changes in our field (and our society!) now, we can change the expected worldview of generations to come.

So, how did we get here? In this chapter, we will briefly (very briefly) review how we moved from a pre-computer, to an early computer, and to our current computer world. We include this chapter to provide some context, but we certainly do not cover all the interesting history around UX that exists. We suggest that you read through the sources we provide—understanding how we got here helps us know where to go.

Before Computers

The word "computer" today is usually used to mean a desktop computer. But, technically, a lot of things are computers—laptops, tablets, smartphones, smartwatches, calculators. Anything that *computes* is a computer. The word actually originally referred to *people* who *computed*. Before there were high-powered, programmable computers to solve complex equations, people would complete long math problems by hand, and so were called "computers."

The job of computer was often held by women, specifically Black women. Today we may call them mathematicians, and their job was to support research like that which was being conducted at NASA.[1] (If you read the book or saw the movie *Hidden Figures*, you are probably familiar with some of these amazing women, including Katherine Johnson and Dorothy Vaughan.) Black women also played a large role in programming the very first computer, ENIAC, built

DOI: 10.4324/9781003181750-4

for the US Army.[2] Today, however, few women complete degrees in computer science. Studies suggest that this is due to myriad factors, including computer and video game marketing being targeted to boys, as well as gender attitudes and behavior within the computer science community.[3]

In this book, we will use the word "computer" to refer to devices like desktop computers, laptops, smartphones, tablets, websites, and apps. So, when we say "before computers," we are talking about the human experience before the digital machines we are so accustomed to today. Of course, ancient tools were designed to "compute" or measure mass (ancient scales), measure distance (Jacobs' staff, range finders, odometers), measure time (sundials, water clocks), compute numbers (abacus, mesolabio, Antikythera mechanism), and communicate (lighthouses).[4] But, the first *computers*, as we use the word today, were machines like the US Army's ENIAC and IBM's 7090.

These early computers were the size of a room and had no screens, mice, or keyboards. Instead, a punch card system was used to input information and output results. A person would punch a stack of cards by hand, creating whatever equation they were trying to solve or program they were attempting to run. They would then hand the stack to the operator, and she would put the stack in the queue, behind the other jobs. To begin a job, the operator would put the stack into a hopper and push "run." Whatever program was punched in the cards would run (if it worked). This process could take an hour or many hours, depending on the length and complexity of the program. Once completed, the "answer" or output was punched out and delivered by the operator.[5] Today, this same process, which could have taken hours, is equivalent to running a program on your laptop or phone that takes maybe a couple of seconds (usually less).

In the 1950s, 1960s, and 1970s, controls were designed solely for the purpose of operating the machine. Computers weren't designed for people; people were merely meant to become another part of the machine. That is, the experience was not user-centric, and computers were not widely used for communication. Computers were still largely viewed as computation machines, and people were expected to adapt to them. Any controls were designed to operate the machine, not to provide the user with an enjoyable experience. No one had computers in their homes; during this time, computers were located in businesses, government buildings, and universities.

However, some tech-oriented people began to think beyond punch cards—and this is where things that we're familiar with today, such as mice, icons, monitors, and software, started being created. For instance, the mouse was developed by Stanford Research Laboratory (now SRI) employee Doug Engelbart in 1965. It was a cheap replacement for a previous manipulation tool—the light pen—which never really took off because of the awkward angle the user had to constantly hold their arm while using the wand-like pen. In what is now called "the mother of all demos," Engelbart conducted a demo of his mouse (so named because the wire looked like a tail), as well as hypertext, objects on an interface, dynamic file linking, multiple windows, and communication between two people over a network with audio and video. In other words, this demo was a very early look into a human-centered computing experience. While the "mouse" title stuck, "bug," which is what Engelbart called the cursor, wasn't as popular.[67] You can watch "the mother of all demos" here!: www. youtube.com/watch?v=B6rKUf9DWRI

The first computer advertised with a monitor was the Xerox Alto, released in the early 1970s. It came with a keyboard and a three-button mouse as well as an 8 ×10, sideways television-like screen. Beyond helping consumers complete tasks, the Alto was also marketed as a communication device—the first public glimpse into computers as interpersonal and public communication tools. Although it was largely advertised as a personal computer, it was very unlikely that people would have one in their home—it cost $32,000![8]

Formative research continued to be carried out at Xerox PARC. David Canfield Smith coined the term "icon" in his 1975 Stanford doctoral thesis and then went on to work for Xerox. Once there, he popularized the idea of icons as one of the chief designers of Xerox Star, first marketed in 1981. Officially the "Xerox 8010 Star Information System," this computer improved upon Xerox's Alto, adding a two-button mouse and folders and costing about half the price of the Alto.[9] Beyond early computers, many program advancements that we take for granted today began at Xerox in the 1970s and early 1980s. Bravo, the first contemporary Word Processor, and Draw, the first drawing program, for example, were developed there.

This initial research at Xerox PARC encouraged other companies to create more efficient computers and software. Apple began producing similar style computers, including The Lisa and Apple II. (Sketches of The Lisa user interface, as it was being developed, can be seen here: www.folklore.org/ StoryView.py?story=Busy_Being_Born.txt.) Apple was also at the forefront of creating the foundation for the contemporary, "user-friendly" software we're used to now. VisiCalc was the first spreadsheet program and was very similar to software that we rely on today, like Microsoft Excel. The mouse could be used to select a cell, cells would auto recalculate when values were changed, and labels and formulas would be suggested to the user as they typed.[10]

WordStar, created by Seymour Rubinstein and John Robbins Barnaby, was similar to Bravo, but it had a much steeper learning curve due to its complicated interface. However, once users understood the program, it was quite powerful. WordStar was the foundation for immensely popular software today, like Microsoft Word.[11] Programs like WordStar and VisiCalc finally provided a concrete reason for families to purchase personal computers for their homes. WordStar allowed people to type letters and other correspondence, and VisiCalc was incredibly popular for personal budgeting. Files could be easily created, formatted, stored, and edited.[12]

These early software were the first WYSIWYG programs (fondly pronounced wizzywig), or "what you see is what you get." The idea behind WYSIWYG software is that the user sees no source code, just the end product. In other words, what the user sees is what the product would look like if it were to be printed out on a piece of paper. This was a huge change and what allowed for the eventual move to massively adopt digital computing technologies, the digital tools with which we are familiar today. When the program no longer requires users to have deep knowledge of programming languages, computers can finally make their move into households (as long as the hardware is affordable!).

The WYSIWYG ideal also led to users becoming further and further removed from fully understanding the mechanisms behind the digital infrastructures that they trust and fold into their daily lives. WYSIWYG technologies are often referred to as "user-friendly." What this really means is that users (those assumed to not be experts) need increasingly less knowledge of how digital programs, websites, and apps actually function. In other words, most current UX research does not focus on the background processes that are making apps and websites function the way we expect. Often, in fact, pleasurable user experiences include *not* having to think at all about how an app or a website works. Many argue that users have become so far removed from technologies' inner workings that they are easily misled, exploited, and taken advantage of, as we explore in Chapter 2. In this book, we hope to inspire you to look deeper into the digital world and to help your participants and users become more digitally literate and critical as well.

Moving Toward a More User-Centric Experience

In 1984, the now iconic *1984* Apple Macintosh ad ran during the Super Bowl. The ad was based on George Orwell's novel of the same name set in a dystopian society where people are meticulously controlled by the "thought police," led by "Big Brother." Apple uses Orwell's premise as a parallel to the computing world, implying that computers are led by companies and not by people. The ad ends with this famous tagline: "On January 24th, Apple Computer will introduce Macintosh. And you'll see why 1984 won't be like "'1984.'" The actual Macintosh computer is never shown in the commercial, but this machine improved upon the technologies that were a part of The Lisa and Apple II, including a handle for easy transport.[13]

Clearly, we can see the move to more user-centric design. Even in the mid-1980s, Apple was concerned with convincing users that they would provide an experience unlike other companies—creating a computer that was actually made for general users to be free, not to be controlled by their technologies' technical whims.[14] In the late 1990s, Apple introduced their now famous motto "Think Different," again attempting to communicate their focus on users who were not the "normal" (technically expert) computer users, tapping into a creative, not necessarily "techie" demographic.[15]

This motto was extremely successful for Apple; it helped them to sell their iMac as well as their later, even more personal devices—the iPod, iPhone, and iPad. These devices, and their related operating systems and programs, clearly show the move into user-centric experiences. Apple's user experience research is focused on making sure that users enjoy not only the product itself but also the brand more generally. Other computing companies have done the same. Microsoft is, of course, famous for their Office Suite—Word, Excel, Power-Point, etc.—as well as, more recently, their Surface laptop. Companies like Google provide not only user-centric programs like their search and Gmail but also devices like their Pixel smartphone line.

As we move through the timeline from the 1990s into the 2000s and up to now, there is a clear shift from people *operating* a computing machine by pressing the right buttons to *using* applications and tools. A few thinkers really guided this process. In his 1996 essay "Software Design Manifesto," Mitchell Kapor (developer of Lotus 1–2–3, another early and popular spreadsheet program) describes a process of design that includes thinking about how to make software not only functional but also usable and delightful. Here, there is a clear linkage between engineering and design. Of course, the product should do what it says it is going to do, with no bugs, but it should also be a pleasure to use.[16]

Perhaps the most famous thinker in the UX world is Don Norman. While working at Apple in 1993, Norman coined the term "user experience." He recognized that the rise of personal computing in the 1980s called for a more experience-based approach that needed to be quite different from the then-common business focus. For instance, in personal computing, the user and buyer are the same person. So, the buying experience needs to be considered, alongside the usability of the actual hardware and software. In the 1990s, Norman also realized that the burgeoning internet and world wide web presented new challenges and opportunities in UX. In particular, he noted how important the experience of a brand was on a website's shop, since the user needed to make it all the way through the content, decide to click "add to cart," and move through the checkout process.[17] Go here to watch Don Norman explain the term "user experience": www.nngroup.com/videos/don-norman-term-ux/

Computing and User Experience Today

While early computing was about operating a machine and performing a task, today computing is about having an experience, no matter if we are learning, working, or communicating. Instead of being the operators of machines, we live alongside them. We not only expect them to provide pleasurable experiences, but also expect that we can easily fold them into our everyday experiences. They become second-nature utilities.

Today, it is the job of UX researchers to ensure that people have a pleasurable experience with brands and products, with as few frustrations as possible. UX researchers start off by asking users about their needs and goals for a particular product and by observing their behaviors in already-existing digital spaces, paying particular attention to "pain points"—parts of the process that produce frustration or annoyance. Later, once a new product (or a new feature on an existing product) is created, researchers run tests with real users, to check that the experience of interacting with the product is indeed as pleasurable and relatively pain-free as the designers imagined.

For Media and Communication scholars, the goals in conducting research are more similar to UX researchers than you may have previously thought. Media and Communication scholars are dedicated to better understanding the complex interactions that occur between people, with the help of, and guided by, digital technologies. Investigating how people experience their digital worlds *in context*—watching them use the apps and websites, if possible, in a real-world setting—is the first step. Understanding how a digital space functions—that is, viewing structural elements and features not as things that get in the way but as parts of the user experience that help complete the story—allows researchers to produce even more salient research in the Media and Communication fields. Understanding how a digital space functions also assists with conducting studies that are diverse and inclusive—not everyone experiences apps and websites similarly. And certainly not everyone experiences them in the same way as the owners and designers of the products!

Typically, the main difference between UX researchers working in tech and academic researchers studying tech is the ultimate purpose of the research: applied versus academic/basic focus. In industry, UX researchers do applied research, with the purpose of making specific recommendations for a design or trying to solve a problem (typically around a particular digital product, such as a website, app, or device, that the company makes). In academia, researchers working in Media and Communication Studies with a UX focus do research to understand how different users interact with technologies (and each other) and how cultural and societal factors come into play in these interactions. Studies like this contribute to a body of knowledge about human-computer interaction, but rarely do academics make specific recommendations or provide solutions. However, as you'll see throughout this book, the research process in both research "tracks" overlaps significantly—and both can learn from each other!

Notes

1. "Katherine G. Johnson," *NASA*, May 25, 2017, www.nasa.gov/feature/katherine-g-johnson.
2. "ENIAC Programmers Project," *ENIAC Programmers*, 2021, http://eniacprogrammers.org/.
3. Eileen D. Bunderson and Mary Elizabeth Christensen, "An Analysis of Retention Problems for Female Students in University Computer Science Programs," *Journal of Research on Computing in Education* 28, no. 1 (1995): 1–18.
4. Cesare Rossi, Flavio Russo, and Ferruccio Russo, *Ancient Engineers & Inventions* (Dordrecht: Springer, 2009).
5. Steven Lubar, " 'Do Not Fold, Spindle or Mutilate': A Cultural History of the Punch Card," *Journal of American Culture* 15 (1992).
6. Brad A. Myers, "A Brief History of Human-Computer Interaction Technology," *Interactions* 5, no. 2 (1998).
7. William K. English, Douglas C. Engelbart, and Melvyn L. Berman, "Display-Selection Techniques for Text Manipulation," *IEEE Transactions on Human Factors in Electronics* 1 (1967).

8. Steven K. Roberts, "The Xerox Alto Computer," *BYTE Magazine*, September 1981, https://archive.org/details/byte-magazine-1981-09/page/n59/mode/2up.

9. "Xerox 8010 Star Information System," *National Museum of American History*, accessed July 18, 2021, https://archive.org/details/byte-magazine-1981-09/page/n59/mode/2up.

10. Owen W. Linzmayer, *Apple Confidential 2.0: The Definitive History of the World's Most Colorful Company* (San Francisco: No Starch Press, 2004), 14–15.

11. Thomas J. Bergin, "The Origins of Word Processing Software for Personal Computers," *PC Software: Word Processing for Everyone* (2006): 32.

12. Linzmayer, *Apple Confidential 2.0*.

13. Ibid., 109–14.

14. Ibid.

15. Rob Siltanen, "The Real Story Behind Apple's 'Think Different' Campaign," *Forbes*, December 14, 2011, www.forbes.com/sites/onmarketing/2011/12/14/the-real-story-behind-apples-think-different-campaign/?sh=41ac8a6362ab.

16. Mitchell Kapor, "A Software Design Manifesto," in *Bringing Design to Software*, ed. Terry Winograd (New York City: ACM Press, 1996).

17. Jakob Nielsen, "A 100-Year View of User Experience," *Nielsen Norman Group*, December 24, 2017, www.nngroup.com/articles/100-years-ux/.

4 Interfaces and Navigation

In 1983, *Time Magazine* switched up its usual Person of the Year: instead of naming a person, the magazine named the computer the "Machine of the Year" and stated, "The Computer Moves In." As discussed in Chapter 3, computers were certainly around before 1983, but the idea that computers were made with the human experience in mind, especially how we view them as social utilities today, began in the 1980s.

Key components for the idea of user experience lie in interfaces and navigation. In this chapter, we first outline what interfaces are, breaking them down into three main types—user interfaces, Advertiser APIs, and Developer APIs. We then provide brief definitions of a list of popular navigation tools. Although as a UX researcher you are not designing interfaces and navigational tools per se, these basics are foundational knowledge for conducting effective user experience studies of apps and websites.

Understanding interfaces and navigation is also useful if you are conducting research in academic fields like Media and Communication Studies. Often, when you are attempting to understand any space in which, or through which, people communicate, it is important to understand the context. How digital spaces like apps and websites are designed (as discussed in Chapter 2) provides specific scripts, constraints, and limitations for users to work in. While it may at first seem that groups of people just "are" a certain way online, it could actually be, and is likely, the case that something about the interface is priming or prompting them to act in ways that reify stereotypes. Being able to speak the language of human-computer interaction as well as recognizing the functionalities of digital spaces is the first step to conducting critical and inclusive studies that help us better understand digital communication phenomena.

Interfaces

An interface is where two different systems meet and communicate with one another. Seismologists study the interfaces between tectonic plates and how vibration is communicated between them. Physicists working in optics consider the interfaces between the material of a lens and the air and how light is communicated between them. As a UX researcher, you will consider the

DOI: 10.4324/9781003181750-5

interfaces between human users and the systems (applications) they are using, and how information is communicated between them.

As a verb, "to interface" is to communicate across this boundary. At the core, this is what a computer interface does: it allows two or more people or things to *interface* through some connection and interaction between hardware, software, and users. Users communicate with the software and then the software communicates with the hardware and perhaps some other software. A good example might be a user's interaction with their email. In order to access an email message, a user might interface with an email application via a web browser, then that application might, in turn, interface with a separate database to access the content of the message. Here, a user communicates with different types of software and hardware in one interaction.

While we often think of humans interfacing with technology, computers also interface with other computers with little or no outside human interaction. In some cases, these computers are embedded into devices, or even into animals (think microchips) and people, in order to allow them to communicate with each other. Today this network has come to be known as the "internet of things" (IoT).[1] The IoT consists of smart devices that are programmed to collect and share data. This collection and sharing do not require any direct, conscious, human-to-computer, or human-to-human interaction. For instance, a person may have a heart monitor implanted that automatically sends data to their doctor without any human intervention, or a piece of manufacturing machinery might have sensors that report its status or location to other machines.[2]

Hardware interfaces include components like plugs, sockets, and cables. You may be familiar with a USB socket or an ethernet cable. The USB socket is a hardware component that allows your flash drive, for example, to *interface* with your computer's operating system. Software interfaces are things like languages, codes, and operating systems, including Windows, Mac, and Linux. User interfaces include pieces like keyboards, mice, commands, and menus. A user *interfaces* with a mouse by moving it around a screen and clicking on the desired options. This then triggers some other action to happen, like opening a file or program, showing that the software and hardware are also *interfacing* with one another. All interfaces have some structure that includes both how the data move through them and an implied function that links to what that interface actually does.[3]

Interfaces that have been made specifically for the user, to allow for and enhance visual user experiences, are called graphical user interfaces (GUIs). GUIs are the visual representations of the interface between user and computer. An example of this are the menus and icons that make up a computer's desktop. The user interacts with these components in order to communicate with the operating system and to perform actions on the desktop.[4] Prior to the development of GUIs, users could still perform most of the same tasks but would interface instead via the command line interface, or CLI. This is still common in the domains of system administration and software development, where little is gained by adding the computing overhead of a GUI. Instead, in these fields, the end goal is to have the system or software manage itself, and since computers can interface more easily with text commands than pointing devices and clicks, a GUI is not only unnecessary but often unwanted.

In addition to GUIs are voice-controlled interfaces (VUIs) and gesture-based interfaces. VUIs, as the name suggests, are controlled by voice commands instead of other human input like a keyboard stroke or a mouse click. Popular VUIs include Siri and Alexa. By saying the "wake command" ("Hey Siri" or "Alexa") to the Apple or Amazon device, the program "listens" and attempts to perform the task for which the user has asked.

Gesture-based interfaces are most prevalent in virtual reality (VR) and augmented reality (AR), where a user performs certain body movements to elicit the connected action.[5] A user playing a VR video game might slash with their empty hands in order to swing a sword in the game. Everyday items are also beginning to include gesture-based interfaces. For instance, in an effort to promote safe driving, BMW has included gesture-based interfaces in their cars. If a driver takes one finger and spins it around, the volume of the radio or GPS will turn up and down. Or, if the driver puts up two fingers, like a peace sign, the touch screen turns on and off.[6]

Usable, Pleasurable, and Accessible

As Don Norman describes in his essay "Emotion and Design," there was much tension between usability and pleasure in the 1980s. Early computer displays, for instance, were black and white. When color was first introduced (first in the 1950s, but really becoming the standard only in the 2000s), no practical reason was cited. Instead color was seen as a superfluous, decorative addition to screens. Some used color to highlight text, but generally color was not seen as a utility. However, companies still rushed to get color displays. Even Norman wouldn't give it up after having tried it; he claims that he felt he *emotionally needed* the color displays.[7]

Usable does not mean pleasurable. As we learned in Chapter 3, just because something works well does not mean it is enjoyable to use. The idea of usability is rooted in psychology, computer science, human factors, and engineering. But, what about the areas that are not normally addressed through science and engineering fields? Here, Norman is specifically pointing to the now-popular field of UX—thinking about the user's entire experience, not just about if the product does what it is supposed to do.[8] Of course, an incredibly important piece left out of the conversation is how color impacts a diverse set of users. Colors beyond black and white on displays can help people with certain visual disabilities navigate these displays easier. In addition, certain colors or color combinations may provide neurodivergent users with increased usability.

Just the opposite is true as well. Attempting to design an app or a website and being too focused on the aesthetic value and not focused on usability obviously leads to a product that just doesn't work. This never leads to a positive user experience. But, sometimes too much color or too many images, in an attempt to be aesthetically pleasing, can also prohibit users with certain disabilities from having a pleasant experience. This is why it is so important that apps and websites include functionalities like the ability to change colors and contrasts.

Norman argues that, when in stressful situations, the best design is human-centered and affective (connected to emotions). If the user feels like they are in a pleasant situation, they are much more likely to be tolerant of minor difficulties. In a sense, it is as if the pleasant, enjoyable product is priming the user to be more accepting in case small, frustrating issues arrive. As Norman says, "attractive things work better."[9] However, we want to be sure that we put a critical eye to this broad statement—attractive to whom, exactly? A stereotypically "ugly" GUI may be perfectly crafted for a group of people with specific needs or cultural expectations.

GUIs

GUIs arguably were first invented at the Xerox Palo Alto Research Center. In 1977 a team designed Xerox Star—software designed to run on a series of personal computers—but the interface was much too slow and was not commercially successful. It just so happened, however, that Steve Jobs visited Xerox during this time and saw Xerox Star. Jobs went back to Apple, hired the original designers of Xerox Star, and created Apple Lisa. This product was also not successful. But, in 1984, Apple developed the successful Apple Macintosh, still based on the general Xerox Star vision. Mac's GUI set the tone for the look and feel of GUIs used today.[10] (We cover a bit more computer history in Chapter 3.)

The Apple Macintosh was advertised as the computer "for the rest of us." The GUI included menus, icons, point-and-click, and mouse-driven processing. It also limited users to contextually correct answers—this means that, once a user makes a selection, the menu limits what can happen next, ensuring that the options are based on the previous selection.[11] For example, if in the main menu the user selects "volume options," only sub-options relating to volume will be visible. This type of technology seems quite fundamental today, and it is likely you don't even notice the behind-the-scenes process. But this type of design is crucial to usability and good experience—and has not been the standard for that long!

Today, GUIs are best made when keeping three criteria in mind:

1. Amount of info—abbreviations should be used well. Wording should be concise and unnecessary details should be avoided. Keep to familiar

formats (e.g., the official way to write a US address or phone number) and keep content tabular with headings.

2. Grouping info—make grouped information obvious through color schemes, boundaries, and highlighting.

3. Information sequencing—there are a myriad of ways to organize information, so part of the UX researcher's job is to explore which order provides the best experience. Depending on the product, some examples include sequence of use, conventional usage, importance, frequency of use, generality versus specificity, alphabetical, and chronological.[12]

Types of Interfaces

When understanding GUIs, it is important to realize that there are different types of users. We often employ the word "user" to describe the more formal category of "end-user." End-users are the everyday people that are expected to use the product. These outward facing apps and websites are officially labeled end-user interfaces, or EUIs.

EUIs are generally all that most users of apps and websites see. If you open an app on your phone or go to the main URL for any site, you are accessing the EUI. It is deceptively easy to imagine that all that exists is the EUI, largely because it is all most users directly experience and because most users seemingly have no reason to view, or even really know about, other GUIs.

Although most people use the word "user" in place of end-user, there are two other main types of users—advertisers and developers. The interfaces for these two groups are usually referred to as application programming interfaces, or APIs. This is because APIs are interfaces that provide slightly more backend access to apps and websites. APIs are where certain types of content are created for end-users. This content will become part of EUIs through components like targeted content and third-party apps.

Advertiser APIs are commonly referred to as dashboards. These dashboards are generally accessible to anyone—as an end user, it is possible to open an advertiser API and see what advertisers, seeking to create targeted content on apps and websites, use. Advertiser APIs are specifically designed to provide pleasurable experiences for those creating ads. Figure 4.1 shows a screenshot of Google Ads' dashboard.

Developer APIs are areas for third-party developers to utilize components of an app to create their own program. For example, the Facebook developer API allows third-party programmers to tap into the data they make available to software developers to create apps and games. You most likely have encountered a login page that asks you to log in with your Facebook or Google username and password. This is an example of that app or website utilizing Facebook's and Google's APIs. Instead of the smaller app having to program the complicated, and, of course, important, the task of setting up a secure, encrypted connection, the programmers leverage Facebook's and Google's expertise and well-tested code. Of course, Facebook and Google get something out of the deal too—the

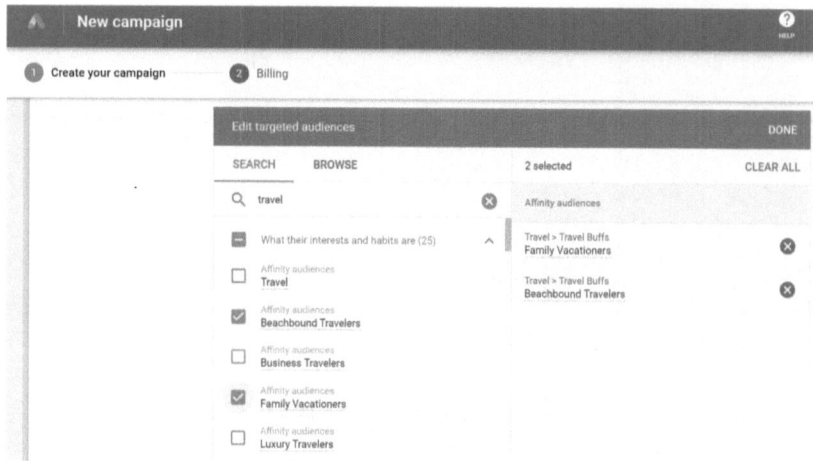

Figure 4.1 A screenshot of the Google Ads' dashboard.

companies are now privy to data created through this connection which feeds back into their model and helps increase profits.

Although popular UX research discussions are aimed at end-user experiences, UX research is also conducted to ensure that advertisers and developers are provided with spaces that are usable and pleasurable. While developers, and even advertisers, often use lower-level interfaces to interact with APIs, software platforms have recently made available GUIs in the form of dashboards for developers. These dashboards often allow developers to see what parts of the API they are using and turn functionality on and off. An example of Twitter's developer API dashboard can be seen in Figure 4.2.

Navigation

An important piece of GUIs is the ways that a user can navigate through your app or website. Knowing the different tools available to provide users with great navigation is useful in UX research, even if actually designing them usually falls under the job of a UX designer or engineer. This section will introduce you to a short list of popular navigational tools.

Breadcrumbs are named as such because they are like the breadcrumbs you drop to leave a trail so that you can find your way back. Breadcrumbs are visual representations, usually of simple page names, that let you know where you are at any time on a website (and sometimes in apps too). Breadcrumbs certainly help with findability, but also they help users to not feel lost; they can always get back to a previous tier or remember why they are on the page they are on.

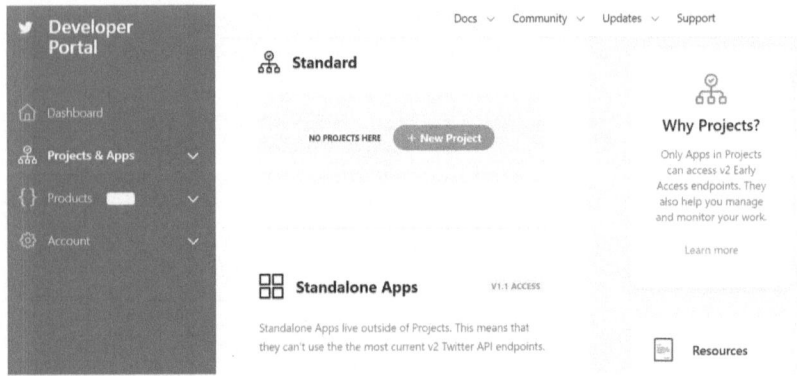

Figure 4.2 A screenshot of Twitter's developer dashboard.

Books   Advanced Search   New Releases   Best Sellers

Books › Cookbooks, Food & Wine › Cooking Methods

Figure 4.3 Amazon's breadcrumbs.

Figure 4.3 shows a screenshot of Amazon's app EUI. At the bottom, you can see the breadcrumbs provided; it is letting the user know that they are in the "Cooking Methods" section of the "Cookbooks, Food, and Wine" category. More broadly, the user is in the "Books" section of Amazon.

Tooltips are brief descriptions that explain what a tool or button does. Tooltips are always hidden on the default GUI and can be seen by hovering over the tool or clicking/tapping. Tooltips save a lot of room and clutter because the wordy instructions are hidden. The hidden tooltips are also nice for users who don't need them; users don't want to constantly see the wordy instructions when they don't need them. However, new users, and users that have forgotten what certain functionalities do, can use the help. Figure 4.4 shows a screenshot of a tooltip in Google Drive. In the search bar, there is a small icon that looks like some sort of generic settings. But, when the user hovers, they get the description "Search options," reminding the user what will happen if they click that icon.

Coach marks are like tutorials that individually highlight certain functionalities that the app or website think will be useful. Often, coach marks will appear when you have opened an app for the first time, when a new tool is

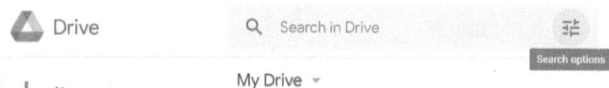

Figure 4.4 Google Drive's search bar filter tooltip.

Figure 4.5 TED's coach mark highlighting casting capabilities.

added, or when you haven't logged into a site for some time. Figure 4.5 is a screenshot of TED Talks' app. Upon opening the app for the first time, a coach mark points out to the user that if they click the highlighted icon in the upper-right corner, they can cast the video to their TV.

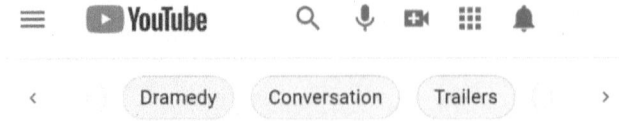

Figure 4.6 YouTube's slider suggesting content.

Sliders are lists that can move horizontally and that provide multiple options. Sliders allow users to see many options while saving space and decreasing clutter on a page. In addition, a slider acts as a menu, while the user stays on their current page. In Figure 4.6, a slider is shown on YouTube with suggested video content categories.

Popovers are similar to pop-ups (external ads or other content that "pop-up" or appear on top of the web content a user is trying to view) but are built into an app or a website. When a popover appears, the content the user had been viewing is still visible behind the popover. Popovers are used to get a user's attention, providing them with something you want them to see or do, over and above what they're currently browsing. For example, in Figure 4.7, a screenshot of the babysitting app Bambino, a popover asks if the user wants to log in to Facebook for a more personalized experience. Popovers are usually easily dismissed by simply clicking/tapping outside of the popover box. As seen in Figure 4.8, a screenshot from TED, popovers can also be controls. As with the first example, the content can still be seen in the background. In this example, however, the popover controls are themselves also partially transparent in an effort to reduce interfering with the video as much as possible.

Sidebars are hidden menus that can slide out and quickly show the user a list of options that otherwise would add clutter and be less aesthetically pleasing. In Figure 4.9, you can see eBay's app when a user first logs in. Typical menu options like "Saved," "Buy Again," and "Purchases" are seemingly missing. However, when the user clicks/taps on the three horizontal lines—fondly known as the "hamburger" because the icon looks like a crudely drawn burger—the sidebar slides out and provides easy access to multiple options.

Of course, there are many more navigational tools, and increasingly more are being introduced as the world goes more and more digital. These is just a small sampling of popular examples, but hopefully they begin to get you thinking about how navigation and interfaces are key to user experience. We hope that being aware of these different design components can lead you to conduct effective studies that are also inclusive, keeping users with different backgrounds and abilities in mind.

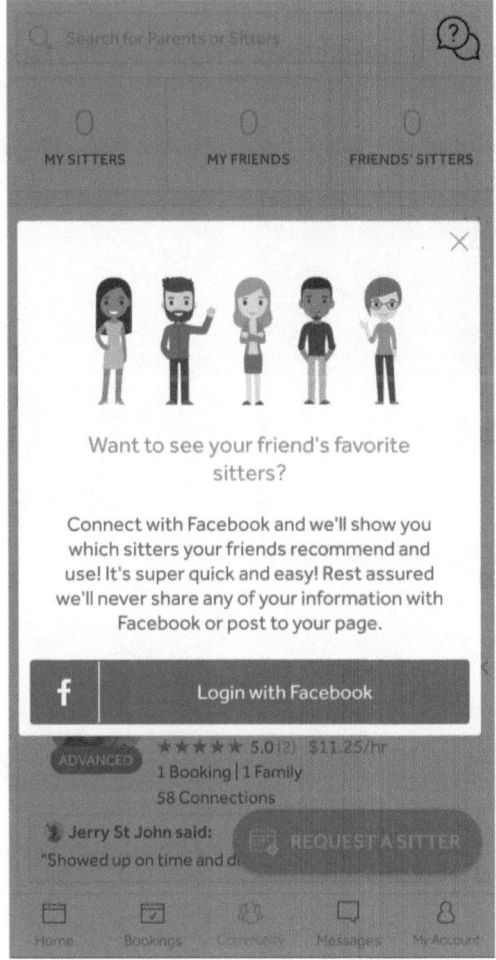

Figure 4.7 A Bambino popover suggesting a Facebook login.

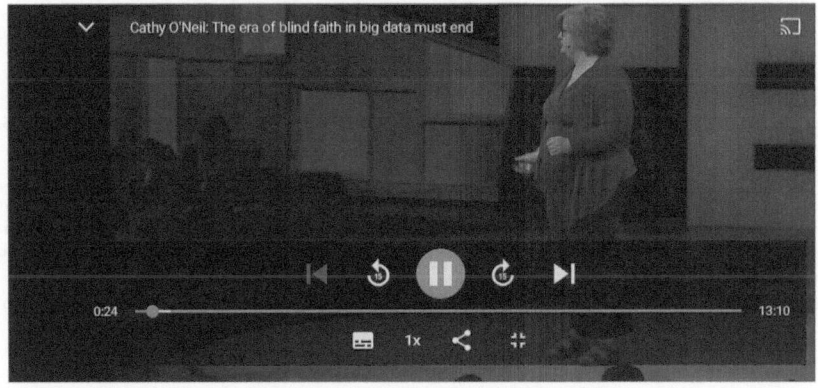

Figure 4.8 A transparent popover in TED showing video controls.

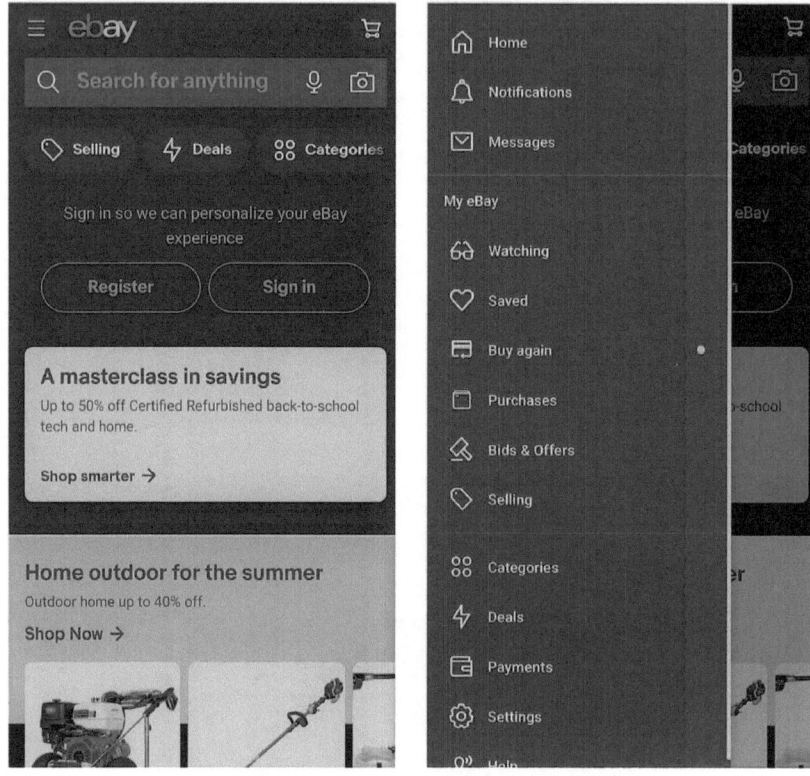

Figure 4.9 The eBay app's initial login view before and after tapping the "hamburger" and revealing the sidebar.

Notes

1. "Interface," *PCMag*, accessed July 18, 2021, www.pcmag.com/encyclopedia/term/interface.
2. "Internet of Things (IoT)," *IoT Agenda*, accessed July 18, 2021, https://internetof thingsagenda.techtarget.com/definition/Internet-of-Things-IoT.
3. "Interface."
4. "What Is User Interface Design?" *Interaction Design Foundation*, accessed July 18, 2021, www.interaction-design.org/literature/topics/ui-design.
5. "What Is User Interface Design?"
6. Matt Burns, "BMW's Magical Gesture Control Finally Makes Sense as Touch-screens Take Over Cars," November 4, 2019, https://techcrunch.com/2019/11/04/bmws-magical-gesture-control-finally-makes-sense-as-touchscreens-take-over-cars/?guccounter=1.
7. Don Norman, "Emotion & Design: Attractive Things Work Better," *Interactions* 9, no. 4 (2002).

8. Ibid.

9. Ibid.

10. Bernard J. Jansen, "The Graphical User Interface," *ACM SIGCHI Bulletin* 30, no. 4 (1998).

11. Ibid.

12. Ibid.

Section Two

Preparing Your Study

5 The Design Thinking Mindset

The best studies follow a particular method to ensure rigor and valid results. Perhaps you have already learned about the scientific method: a linear process that begins with a question, dives into a specific research area, hypothesizes what the findings may be, tests those hypotheses with experiments, analyzes the data, interprets the findings, and reports the results. The scientific method is most often taught in the lab sciences, is viewed as objective, and should produce results that are replicable and generalizable.

You may have also learned about more qualitative research that follows a similar, linear pattern. But, this type of research typically starts with an open research question instead of hypotheses and employs more "qualitative" methods that are not experiments but still gather data—interviews, observations, surveys, focus groups, and so on. These "pseudo-scientific" studies are sometimes viewed as less objective, less generalizable, and less replicable. Yet, they are still rigorous, valid, and highly valued in academic communities for understanding cultural norms, smaller groups of people, and snapshots in time.

In the hard sciences and academia alike, it isn't so much that one type of research method is *better*, but which is more fitting for the context, the type of questions you want to answer, and the types of data you can collect. In fact, the scientific, "objective" method has been used to "prove" many racist, sexist, and biased theories including hard "scientific" studies that were used to support white and male superiority. It is generally argued that scientific research often takes on dominant viewpoints, sometimes without the researchers even realizing.[1]

What "science" and "scientific research" are is a timely topic as well. Since the early 2000s, there has been an ongoing "replication crisis." Meta-studies have shown that the majority of scientific studies are impossible to reproduce. This is mainly an issue in the social sciences and medicine,[2] but studies have also shown it to be a problem in the natural sciences.[3] Thus, our goals in this textbook do not include attempting to fit into some "traditional" definition of scientific research. Instead, we adopt the Design Thinking Mindset to conduct rigorous and relevant research that considers people with different experiences and backgrounds. The Design Thinking Mindset takes an empathetic approach

DOI: 10.4324/9781003181750-7

to suggesting data-driven changes to apps and websites, but it is also a helpful, innovative process for thinking through academic research studies.

The Design Thinking Mindset

Design Thinking is, at the core, a way of creative problem solving. The Design Thinking Mindset is a guide to understanding poorly designed features and functionalities, as well as unknown issues in products, by studying human-centric solutions that focus on what is most important to users.[4] It is essentially a framework to ensure that UX research is rigorous and valid. Importantly, the Design Thinking Mindset is different from a traditional research mindset because its main goal is to be human-centered while still being business- and technology-centered. Focusing on all these (sometimes competing) objectives can certainly prove challenging, especially when companies have deadlines and budgets to meet and stakeholders to please. But, the Design Thinking Mindset allows for solid research to still be carried out, even in the most cut-throat of corporate environments. This is not to say that all UX research, even that which follows the Design Thinking Method, is diverse, empathetic, and unbiased. However, Design Thinking is a good starting point for completing quality research that can, in fact, be inclusive.

As defined by IDEO, a global design company, the Design Thinking Mindset is

> a human-centered approach to innovation that draws from the designer's toolkit to integrate the needs of people, the possibilities of technology, and the requirements for business success.[5]

Let's break that down. First, a human-centered approach is one that takes into consideration diverse users, varying usage contexts, and inclusive end-user interfaces. For our purposes, we may consider this a *user*-centered approach. But, we want to be sure to not lose the human aspect; only viewing users as objects to be studied is problematic because it is easy to fall into the trap of thinking of users as robots that use products similarly to each other and in a vacuum. Users should be thought of as holistic humans—with different contexts, different experiences, and different personalities—all which shape their user experience.

As a UX researcher, it is important to remember that you are *not* the user! Your experiences cannot be considered representative or generalizable. Instead, the design choices you make should be based on what you learn from users. Being human-centered really includes two main things: (1) You design for the actual people that use, could use, or you would like to use your product, and (2) you conduct studies and suggest changes that take into consideration diverse and intersectional identities.

Second, drawing from a toolkit to combine what users need and what technologies can actually do, is essential. Here we are talking about keeping on brand and not suggesting changes or conducting studies that are not in line with the app's or website's mission or that are not feasible from an engineering standpoint. There is nothing worse than a UX researcher who isn't aware of the complete process—not just the research they are conducting but the overall capabilities of the product and what the UX designers, engineers, programmers, and other members of the team have as goals, pressures, and constraints.

As for the third part of that definition—business success—even though you are the UX researcher, that doesn't mean you can completely step out of a business mindset. Everything you do must, in some way, consider the success of the app or website. Of course, success can be defined in a myriad of ways: Is the product publicly traded? Privately owned? Non-profit? Are there specific timelines? Expectations? Goals to be reached? Initiatives to focus on? When you begin working for a company or on a product, it is important to quickly learn what type of work is needed that can both provide users with better experiences and help the business remain, or become, successful.

These three parts of the Design Thinking Mindset definition are often referred to as desirability, feasibility, and viability (note how these categories overlap with the seven facets of good UX in Morville's honeycomb from Chapter 1). Products must be desirable to users, but they must also be technologically feasible and economically viable.[6] Tim Brown, a Design Thinking pioneer, explains this way of thinking in his TED talk. You can watch the video here: www.youtube.com/watch?v=UAinLaT42xY

Five Stages of Design Thinking

This textbook is loosely structured on the five stages of Design Thinking, guided by the Hasso Plattner Institute of Design at Stanford University: Empathize, Define, Ideate, Prototype, and Test.[7] Even though these five stages (or phases) are presented as linear here, it's important to note that design thinking should ideally be an iterative, creative process. This means that in reality, sometimes two or more stages of the process might be happening simultaneously, and a product can move between the five stages in different directions at different times of its development.

Empathize

UX studies typically begin with the Empathize stage which is all about understanding the user. As human-centric researchers, we should always be concerned with who users are, what they care about, and what they need. The Empathize stage is yet another reminder that, as a researcher, you are *not* the user! This revelation can be especially difficult when you are studying an app or a website that you use every day or have a lot of experience with. It is very easy to rely on your own experiences or stories from friends when making choices in the research process. But these are not data, just anecdotal evidence that is not really representative of all people. Thus, to empathize is to deeply understand who your users are and what they care about.

This first stage in the Design Thinking process relies on your ability to engage with users, discovering their thoughts and values. Of course, these thoughts and values may not look like yours; they may even directly clash with how you view the world and live your life. But, to empathize isn't to agree, just to understand. It isn't your role as a researcher to force your own thoughts or only attempt to understand those with whom you agree or share similar experiences.

During the Empathize stage, you use methods that allow you to observe, engage, watch, and listen. The goal is to let users be in their own context as much as possible, sharing stories and experiences with you that provide insight into how they use the technology, what they like about it, what they find frustrating, who they are, and what they need, both from the app or website and generally as a human. Methods often used in this stage include emotional journey mapping, contextual inquiry, breakup/love letters, and screenshot diaries.

Empathize stage methods can be used in academic research for studies that aim to understand people's subjective experiences and emotions, particularly when researchers are interested in knowing more about specific groups in specific contexts. In traditional Communication and Media Studies research, Empathize methods are qualitative methods such as interviews and focus groups—methods that focus on attitudes and feelings, rather than behaviors. Research in the Empathize stage is similar to exploratory research in the academic paradigm, in that there typically isn't a very focused research question guiding the research—it is more about exploring what is out there that can lead to a more defined project down the line. Parallel to this, in the Design Thinking process, Empathize is followed by the Define stage.

Define

In the Define stage, you are already aware of the users' needs, desires, and pain points because you have discovered them through some Empathize method(s). Now, you must focus and clarify your findings. The Define stage is particularly important because it is when, as the researcher, you start to the understand the research problem and create a concise and clear Problem Statement. (In the academic research process, this would typically be the first stage of the research process—deciding on the research question or hypothesis to guide your study—unless the study is exploratory.)

A Problem Statement is a brief, one- or two-sentence description of the user experience issue that you will be researching, collecting data about, and, eventually, proposing a solution for. A good Problem Statement focuses on one, small but prominent frustration or misconception that users have with your app or website. For example, a Problem Statement like "Facebook isn't cool anymore" is much too broad and assumes too many things. It is trying to be everything for everyone. Instead, a Problem Statement may be something like "Users can't find Facebook's Most Recent button" or "Users don't think Facebook is doing enough to fight fake news." These statements are clear and small, and not only speak to users' desires but also take into consideration feasibility and viability. Other stakeholders in the company can see how making small changes to the end-user interface and usability upgrades (making a button more visible) or showing the organization is interested in contemporary issues like fighting fake news (by, for example, giving more reporting options) is in line with the company's mission and helps them continue to be a successful business.

A Problem Statement is necessarily guided by the analysis of the data collected from your Empathize method(s), and it is often also informed by previous research that has focused on similar issues, ensuring that you, as the researcher, have the background information and most contemporary methods to best understand the problem. Methods often used in this stage include problem trees and personas. Technically, these are more analysis techniques than research methods in the traditional sense, as they don't involve a collection of information. Typically, in the Define phase, different team members (including designers, engineers, and other stakeholders) will brainstorm the Problem Statement together. Here, not only is it important to consider what you discovered about the needs of users but it's also crucial to talk to other stakeholders (both internal and external to the organization) to make sure their needs and desires for a particular product or feature align with users' needs.

The Define stage as a whole is closest to the research proposal stage in academia, where the researcher conducts a literature review (of other's research findings) to understand what work on the topic was completed previously and to formulate research questions to answer or hypotheses to test. Again, this is the key difference between Research Questions and Problem Statements. Research Questions typically come first in an academic research process, whereas Problem Statements are typically generated from prior Empathize research in UX industry work.

Ideate

The third stage of Design Thinking is Ideate. After creating a Problem Statement and researching some background information, you can focus on generating ideas on how to make the users' experiences more enjoyable and inclusive. You have identified a problem; now it is time to think about solutions. In this stage, the goal is to begin thinking about what changes could be made to an interface or a process. The Design Thinking Mindset challenges researchers to not just do what has always been done or what is easiest. Instead, methods are employed that help researchers to think "outside of the box," dreaming up solutions that speak to the data collected. In the Ideate stage you are not trying to find the one, perfect solution—just a lot of different possibilities!

The Ideate stage is unique because it includes external as well as internal methods. This means that some Ideate methods are completed only by those working within the company (internal)—such as UX researchers, designers, engineers, product managers—while other methods include recruited participants (external). Methods often used in this stage include cognitive mapping, darkside writing, and card sorting.

Ideate methods can be used in academia whenever brainstorming is needed or when academics are working on an applied project, where they provide recommendations. For example, Ideate methods can be used to think through how a particular technology or communication process can be made more inclusive to diverse groups of people. Ideate methods are also great activities to be completed by participants taking part in focus groups, to help researchers observe how people come up with solutions to problems. Ideate methods are also valuable in participatory academic research, where research participants take an active part in the research, brainstorming solutions together with researchers for the ultimate purpose of solving a problem in their communities.

Prototype

In the fourth stage of Design Thinking, Prototype, UX designers typically take over for the researchers. Here the goal is to begin to make new versions of tools, buttons, interfaces, and so on based on the previous three stages—that is, you are actually creating the solution to the problem that you brainstormed in the Ideate phase. A prototype is a mock-up or a replica of what a new product or an altered app or website would look like and how it would function. There are different stages of prototyping, ranging from low-fidelity "rough drafts" that can be as simple as pen-and-paper sketches (these are sometimes called sketches, mock-ups, or wireframes if they show predominantly layout) to high-fidelity, interactive, computer-generated renderings of the UI, built using software like Figma and Adobe XD.

Prototyping does not have a similar process or stage in academic research, largely because academic research is usually not applied research. In other words, it is not typically about creating a tangible product or recommending solutions. However, as with Ideate methods, academic researchers can use prototypes as part of their research activities to better understand how participants would interact with an imagined technology. We believe that attempting to incorporate prototyping activities into Communication and Media Studies would lead to less isolated, more collaborative, and more creative research. For example, researchers could create a prototype of a new social media platform and ask focus group participants how they would interact with it. This type of research gives tangible insights into how participants move through digital spaces and the ways in which tools and functionalities may privilege certain people or certain uses over others.

Test

The final stage of Design Thinking is Test. Testing involves asking participants to engage with the created prototypes, whether it is prototypes of completely new technologies or prototypes of improvements and changes to existing technologies. In this stage, you're checking whether the product meets the needs of the user or what versions of the product provide the best user experience. Testing allows you to isolate pain points, that is, areas of the design that the user has problems with. For instance, maybe navigating to the home page from somewhere else in the website is not intuitive because the "home" button is not easily findable. Methods often used in this stage include A/B testing and usability testing.

Testing methods, when used as part of academic research, are particularly useful for researchers concerned with cultural power imbalances and accessibility. Testing existing technologies, websites, and apps with diverse users can provide empirical evidence for accessibility issues and power imbalances around identity that are baked into technologies. This can help show that digital spaces are often designed for a very narrow range of people with certain characteristics (often able-bodied, white, middle-class men) and are exclusive to others by their very design.

To reiterate an important tenet of Design Thinking: unlike most academic research methods, the Design Thinking process is *not* meant to be linear. Instead, projects should be able to move through the five stages organically, allowing new data and interpretations to guide researchers to the obvious next stage. The researcher can return to any previous stage to which the data lead. The collaborative, iterative, and nonlinear nature of the Design Thinking Mindset is certainly different from the traditional research methods you have probably learned. It straddles the line between inclusive, human-centered research and business-minded and successful product development.

Design Thinking in Action

Let's go through a very simple example of Design Thinking in action. Imagine you work for a successful tech company that has decided to invest in creating a new social media platform. Aside from knowing that you want to create a new product that provides a social media experience, there are no other expectations or directions for this project.

Empathize

As the UX researcher, you would start off with the Empathize stage, asking users about their needs, desires, and pain points. Because you haven't developed your platform yet, you might seek out current engaged users of social media and talk to them about their experiences with and thoughts on social media. You would probably conduct some interviews, do some contextual inquiry (observe how users interact with current platforms and ask them about their use), and map out their emotional journeys when using social media. You would ask questions not only about their current experiences and pain points (what's annoying about current social media apps) but also about an ideal future scenario. What would people ideally *want* from a perfect social media platform? What would make them choose your platform over others that exist in the market?

Define

From your Empathize research, you might discover that people love interacting with their friends on social media but wish they could feel more "present" during the interactions. On the basis of this knowledge, you would create a problem tree, aiming to get to the root of the problem of perceived lack of presence (for more details on problem trees, see Chapter 14). Through your work on the problem tree, you might decide that the cause of the problem is the limitations of the screens that are typically used for social media—they are small and 2D. Here, you would define your Problem Statement as: Social media apps limit the feeling of presence in online interactions because of screen limitations. Now you know what problem you are trying to solve through the development of your new app.

Ideate

In this stage, you would get the team together and brainstorm some ideas on how to overcome the problem. You could, for instance, organize a darkside writing exercise (more on this in Chapter 16), asking team members to come up with the very worst possible social media platform for present interaction (e.g., one that just lets you text with limited characters, no photos, no videos, no calling through the app) and flip the "worst idea" statement for possible solutions. Let's say that at this stage you come up with the idea to incorporate virtual reality (VR) into your social media platform, so that users could feel like they're interacting with their friends in the same space at the same time via the technology.

If following the Design Thinking process linearly, we would move on to Prototype. However, as the researcher you have yet to really explore virtual reality options and how diverse users respond and employ virtual reality functionalities. So, instead of just moving blindly into the Prototype stage, you could loop back to Define, learning more about contemporary virtual reality applications.

Define (Again)

Returning to the Define stage, you don't necessarily need to create a new problem tree or problem statement. However, it is time to conduct a little research about virtual reality in social media spaces. Back in the Define stage, you can research both academic articles and industry blogs and UX pages like nngroup. com and interaction-design.org. Similar to a literature review, you can begin to catalog what has been done before and how other companies have added virtual reality to their social media platforms.

Ideate (Again)

Now that you are better acquainted with virtual reality trends, you can bring this information back to the Ideate stage and brainstorm possible changes that specifically employ virtual reality functionalities, guided by what you have learned from your second Define stage research. For example, you might decide that adding holograms or 3D cartoons of friends on social media would be enough "virtual reality," or you might play around with the idea of sending out free VR goggles to everyone who signs up for your platform. Now that you have some ideas, you are finally ready to prototype!

Prototype

During this stage, the UX designers, working with you as the UX researcher (and your findings), would create some mock-ups of what this social media platform would look like. Obviously creating a prototype of a virtual reality environment is hugely technically complicated. So, start small. Choose one idea from your Ideate findings, such as the idea that your users might enjoy creating 3D cartoons of themselves that could then talk to friends using their phones' microphones instead of just using text or 2D video chat. The UX designers would create a working prototype of this 3D cartoon feature and send it to you to then move to the final stage, Test.

Test

In the Test phase, you are finally ready to see how your users respond to the changes you think best solve the Research Problem. In this example, you may decide to conduct some A/B testing with different size 3D cartoons or with holograms versus 3D cartoons, to see which version possible users would prefer.

As you can see, Design Thinking is not a straight step-by-step process. It allows for adding, taking away, looping back, rethinking, failing, trying again, testing, and innovating. It's flexible and creative and places people right in the middle of any research and design project. Now that you've got the right mindset for human-centered design, you can start planning your research project, which we cover in the next chapter.

Notes

1. e.g., Sandra G. Harding, *The Science Question in Feminism* (Ithaca: Cornell University Press, 1986).
2. Jonathan W. Schooler, "Metascience Could Rescue the 'Replication Crisis'," *Nature News* 515, no. 7525 (2014), https://doi.org/10.1038/515009a.
3. Monya Baker, "1,500 Scientists Lift the Lid on Reproducibility," *Nature News* 533, no. 7604 (2016), www.nature.com/articles/533452a.
4. "Design Thinking," *Interaction Design Foundation*, accessed July 18, 2021, www.interaction-design.org/literature/topics/design-thinking.

5. "Design Thinking Defined," *IDEO Design Thinking*, accessed July 18, 2021, https://designthinking.ideo.com/.

6. "History," *IDEO Design Thinking*, accessed July 18, 2021, https://designthinking.ideo.com/history.

7. *An Introduction to Design Thinking: Process Guide*, Hasso Plattner Institute of Design at Stanford, 2013, https://web.stanford.edu/~mshanks/MichaelShanks/files/509554.pdf.

6 Planning Your Research

Planning your research project ensures that everything runs smoothly and that you get valid insights. This chapter walks you through all the things you need to do before actually starting your study, including formulating a clear research question(s) or objective(s), getting a thorough understanding of background information and previous research, recruiting the right participants, considering ethical issues and accessibility, and ironing out all the logistical details, such as the place and time for conducting a study.

Research Questions or Objectives

Every research project starts with figuring out what you ultimately want to get out of it. What problem are you trying to find a solution for? What question are you asking that you would like answered? What are your objectives for doing this study? Essentially, what information do you want to know at the end of your project that you didn't know before?

Qualitative academic research is guided by Research Questions that the study aims to answer. Research Questions are typically "what," "how," or "why" questions, and qualitative research aims to describe and explain phenomena to add to the body of knowledge in a particular field. UX research is typically driven by Research Objectives or goals to ultimately inform the design of a new website or app or help guide improvements or the creation of new features related to an existing product.

The Research Question or Research Objective guides the choice of research method. Certain methods are better at finding out certain information. For example, if your objective is to understand why teens use social media, interviews would be a good choice because they are an attitudinal research method that uncovers what a participant is thinking. If your objective is to understand whether a website is easily navigable, usability testing or some other form of observation is the appropriate behavioral method that can show you, objectively, how a person moves through a digital space.

DOI: 10.4324/9781003181750-8

Background

Before you start doing your own research, you should first get a thorough understanding of background information related to your topic. In academia, this is the literature review, where you find relevant academic articles and books that have been written by researchers in a particular field on a particular topic. The point of the literature review is to identify a gap in the body of knowledge that you can fill by conducting your study. Perhaps a lot of research has been done about how teens use social media for socialization, but very little work exists on how *seniors* socialize on social media platforms. That's a gap that you can aim to fill!

In industry, when you are working as a UX researcher, you will want to review documents in your company that have to do with the study you're planning (including any previous related research studies), as well as analyzing your competitors and their products, so that you can understand what works and what doesn't. Such a review can also lead to helpful insights into how changes and new features have been implemented. Make sure that you understand how your product or new feature differs from your competition. By thoroughly examining the features and functions of your competitors' products, you can better understand what your product needs to provide a unique and hopefully better user experience. This is called a competitive analysis, and, as well as looking at what is out there currently, it also involves understanding untapped markets, to see where your newly developed product can fit in.

Participants

Participants are the people who take part in your research study. Think carefully about your sample—the people that you need to talk to and observe to get answers to your Research Question or to fulfill your Research Objective. In UX, the participants you recruit for research should be representative of (possess characteristics similar to) the target users for your website or app.

The first thing to consider in deciding on your participants is demographic characteristics. Are you interested in a particular group of people based on age, income level, location, marital status, disability status, etc.? Are you interested in comparing different demographic groups? Or are you trying to generally understand the average person's experience and want to talk to or observe a wide range of demographic categories?

Next, think more broadly about other key criteria for participation in your study. This will vary from project to project. Is it important that you talk only with people who have had some experience with a particular technology or who work in a particular field? If you are interested in a broad target user base (i.e., you're designing something "for everyone"), consider representing both the average or mainstream user and more extreme use cases or outliers. So, if you're designing digital products, consider recruiting people who have day-to-day working experience with computers and also those who are not very digitally literate, to get the full spectrum of possible users.

A note here about diversity in sampling: Think carefully about what is considered mainstream or average. For decades, the "average" user (that digital products were typically designed for) was a white, able-bodied, straight, male. As researchers and designers in the 21st century, we should be actively moving away from such a narrow-minded understanding of the universal person. This means you should always be thinking about diversity in your research, and particularly as you are recruiting participants. Sometimes you may be interested in a very specific group of people (for instance, if you're designing an app for moms or for Latina entrepreneurs), and in that case, you obviously want to do research with your specific target group. But in cases where your target market is broader, you should make sure you recruit a diverse group of people to ask questions and test products on.

Once you have decided on the desired make-up of your sample, you need to think about recruiting (finding and signing up) your participants. Where or how will you find these people to take part in your study? Some recruitment methods include: tapping into your existing networks and asking people you know to spread the word about your research (in UX industry research, researchers often rely on their co-workers to test new products); advertising and promoting your study on social media; putting up flyers in public places, getting the word out into the media with press releases (particularly helpful if you're working on an exciting, impactful study that's newsworthy); using existing email listservs; or reaching out to partner organizations who can help connect you with the people you want to talk to. You can also do intercept or guerilla research, by going to a public place and simply asking strangers to answer a few quick questions. There are also recruitment firms that can do the recruitment for you, if you have the budget to pay them!

Think about building incentives into your budget—you're likely to be more successful in finding participants for your study if you remunerate them for their time. Incentives can range from a cup of coffee (especially in guerilla research) to $20–$50 gift cards for an hour-long interview to being entered into a drawing to win a bigger prize for participation in a focus group. If you are a student, you are unlikely to have much resources for incentives. Get creative (perhaps you can tap into some clubs or student networks you belong to, ask family friends to help, or offer to watch pets or kids in exchange for people taking part in your research), but also don't underestimate people's general willingness to help out with student work. Don't be shy asking everyone you know to help you find participants for your study!

How many people should you ideally have participating in your study? The answer is: as many as possible based on the resources (time, budget, and personnel) you have. Though qualitative research does not require a specific sample size like quantitative research does (because you are not trying to generalize beyond your sample), you should aim for at least five participants per study to get some meaningful insights.

Ethical Considerations

Ethical considerations are a critical part of any research. If you're doing research at a university, you need to get permission to conduct your research from your Institutional Review Board (IRB). An IRB is an internal committee that evaluates research plans to make sure they are ethical. IRBs are less common in industry, so industry researchers rely more on their own moral commitment to ethical conduct.

The golden rule of ethics in research is to cause no harm, whether physical (bodily) harm or psychological harm. While physical harm is easily identified and unlikely to be a major consideration in UX research (unlike biomedical research), psychological harm is trickier. Think carefully about whether the research you're proposing could make your participants uncomfortable in any way. Are the questions you're asking too intrusive, insensitive, or related to past traumas? For instance, if you're working on developing an app for tracking chronic illness, you need to think ahead of time how asking participants about their health conditions might be upsetting and make a plan for working through such difficulties with empathy and respect.

Other very important ethical considerations in any research are anonymity, confidentiality, and privacy. Anonymity means that no identifying information (including name, phone number, etc.) about your participants is collected, so the data collected cannot be linked to a specific person. No in-person research can be anonymous. The most common type of research that is anonymous is surveying, where respondents fill in a questionnaire online about their habits, attitudes, opinions, etc. but do not provide any personal information. Do be aware that indirect identifiers, such as gender and race collected with other information (e.g., personal stories), can jeopardize anonymity. Whenever you collect any identifying information about your participants, the study is no longer anonymous and you have to be careful to keep that data confidential.

Confidentiality means that any information collected during the research process will not be shared beyond the research team. When data are reported and presented publicly, they should be presented in the aggregate (not per individual), or if particular quotes or stories are used as evidence (often the case for qualitative research), they should be anonymized (i.e., no real names or other identifying characteristics should be used).

Whereas confidentiality pertains to storage and sharing of *data*, privacy pertains to the *individual* and their right to have control over if, how, and when their personal information is collected and shared. Privacy is related to consent in research. Whenever there is a reasonable expectation of privacy, you may only collect information from participants (observing, recording, etc.) with their consent. For instance, you cannot record conversations in a coffee shop without getting explicit consent from the people involved.

This is where the notion of "informed consent" in research comes in. Before conducting any research activity, you should get consent from your participants to collect their data. The "informed" part means that you tell them what your study is about (give as much information as possible without affecting the results) and that their participation is voluntary. This means that a participant can decide to leave the study at any point, including *after* data collection has started. Researchers typically provide consent forms for participants to sign before a study is conducted, but you can also collect consent verbally if it is recorded. Remember that you are asking for consent not only to *collect* their information and use it for your desired purposes (e.g., to make recommendations for feature improvements to a product) but also to *record* the interaction. A sample consent form is provided on usabliity.gov (along with a plethora of other templates, resources, and guides) here: www.usability.gov/how-to-and-tools/resources/templates/consent-form-adult.html

Ethics in Practice

You might have read about (or maybe even remember) Facebook's controversial social experiment conducted in 2012. For one week, the company manipulated the algorithms behind the content of its users' News Feeds, showing some people more negative content (by filtering out any positive content shared by each user's friends) and showing other people more positive content. This was all done in an effort to see how the sentiment of a News Feed affects users' moods (as measured through them posting positive or negative status updates after being exposed to the manipulated feeds). The major controversy around this study was that Facebook manipulated the feeds without the users' explicit consent, arguably causing psychological harm by eliciting negative moods (for those who were shown more negative content).

As with much research, providing *all* the information about a study to participants ahead of time is not a good practice, as research relies on understanding how people act naturally (without changing their behavior to please the researcher by acting out "desired" results, for instance). But *some* acknowledgment that users were inadvertently taking part in research, even if in the form of a debrief afterward, would have made the Facebook study much more ethical. Debriefing refers to the process of giving participants more information about the study, after the interview, experiment, focus group, or other research activity took place. This is a crucial part of informed consent, so the participant can truly understand what they participated in and why—this is particularly important when deception was necessary as part of the study design (as was the case with the Facebook social experiment). A debrief also gives participants the space to ask the researcher questions about the study.

For more on Facebook's study, see this Wired article: www.wired.com/2014/06/everything-you-need-to-know-about-facebooks-manipulative-experiment/

Accessibility Considerations

The Americans with Disabilities Act (ADA) of 1990 is a civil rights law that prohibits discrimination based on disability, whether physical or mental disability. This not just covers prohibiting active acts of discrimination (such as not hiring somebody because of their disability) but also ensures equal access to public and commercial spaces to individuals with disabilities. It includes the provision of reasonable accommodations to make physical spaces equally accessible to all (think features such as wheelchair ramps, parking spots designated for those with disabilities, and Braille used to mark elevator buttons).

Like physical spaces, we move through digital spaces. In January 2018, Section 508 of the Rehabilitation Act of 1973 requiring that "federal agencies' electronic and information technology is accessible to people with disabilities, including employees and members of the public" was revised with new standards to ensure equal access to current technologies.[1] Any *federal* website or app is required to comply with Section 508. All websites and apps aren't mandated to comply with the revised Section 508 standards, but many websites are still required to be accessible to all users to comply with state laws, institutional laws, or certain grant requirements. Even if your websites and apps aren't covered by any legal or formal requirements, it is still best practice to ensure your products are accessible equally to all.[2]

What does this mean in practice? Considerations around accessibility include:

- Provide a transcript or closed captioning with multimedia for deaf and hard-of-hearing individuals.
- Do not rely on green and red colors for distinguishing information on your site on account of colorblind individuals.
- Every image should have an alt-text equivalent (a text explanation of the image content) so that screen readers (programs that read websites out loud for visually impaired individuals) can "read" the images, too.
- Blinking and flashing components on a website should be able to be turned off, for those with seizure disorders that can be triggered by flashing images.

Clearly, accessibility is a crucial component of UX *design*, but what does all of this have to do with UX *research*? As an ethical UX researcher, you should be actively thinking about accessibility as you develop your interview guides and research plans for testing. So, for example, during Empathize phase research, make sure to ask your participants questions directly related to accessibility of technology and digital products. And when conducting usability tests, make sure to recruit diverse samples of participants to navigate through your website or app, ensuring that your products are truly usable by all.

Logistics: Space and Time, Remote Versus In-Person

Logistics, such as time and place, are a key consideration for research. Research activities, such as interviews or usability tests, should be held at a convenient time and place for both the participant and the researcher. Make sure to schedule at least 30–60 minutes per participant to have time for introductions and debriefing. You can use any formal setting available to you, such as a classroom, office, or lab, or an informal (often more comfortable) setting, such as a coffee shop or park to conduct research, depending on your resources and specific project needs. Remember to consider how busy and noisy your setting is! As you will likely be recording your session, a quiet space is important for the recording to be useful. You should also consider the privacy of the participants and be aware of who else around you can hear your interaction in a public space, especially if you are discussing sensitive topics.

When thinking about logistics, it is important to consider the pros and cons, as well as the feasibility, of conducting research in-person versus remotely. Almost all research activities, barring certain contextual observational studies that pay a lot of attention to the participants' environment, can be conducted remotely, be it online or over the phone. For instance, you can do interviews using Zoom or run remote-moderated usability tests by having participants share their screens with you as they navigate through your website. Remote studies have the advantage of allowing participants to be recruited without geographical constraints and are also recorded more easily through computer software or phone voice recorders. The disadvantages of remote studies are the lack of nonverbal cues (if you cannot see a participant's face) and other rich contextual cues that can provide interesting research insights. Carefully consider the pros and cons of in-person versus remote testing for every study you do—there's no one-size-fits-all research design!

Now that you've finished planning your research project, you can get started on actually doing the research (the most fun part). But before we get into all the exciting UX methods you can use in your study (see Section 3), we end this section with the next chapter that details how you will ultimately report and present your research findings.

Notes

1. "About the ICT Accessibility 508 Standards and 255 Guidelines," *U.S. Access Board*, accessed July 18, 2021, www.access-board.gov/ict/.
2. "Accessibility Testing for Websites and Software," *Section508.gov*, accessed July 18, 2021, www.section508.gov/test/web-software.

7 Reporting and Presenting Research Findings

After designing your research study, collecting the data, and analyzing the data, you will want to share your fascinating insights with others. Whether you're working in academia or in the UX industry, you need to first and foremost consider your audience for any type of report or presentation. Who are you sharing your findings with and why? Are you sharing insights with your UX team in a workshop format, so that you can brainstorm decisions together about a future design? Are you writing a research paper for your professor as a class project? Are you presenting at a conference, to share the knowledge you have gathered with others working in your field? There are typical reporting (written) and presentation (verbal) formats in both academia and industry, which we outline in detail in this chapter. You should stick to these formats at least loosely (though there is always room for a little creativity, especially when presenting), but always make sure to tweak the specifics so that they make sense to and are appealing to your audience.

Writing an Academic Report

In academia, the findings from research studies are written up in the form of research papers, made up of specific sections that we outline in the following. Research papers are typically published in academic journals. Whereas quantitative research typically presents a lot of numbers as the results, in the form of statistics, qualitative research is more like storytelling, because your findings are based on words and observations. So even though you have to stick to a particular format (academia is much stricter in terms of reporting format than industry), you can still be creative in how you tell your research story. You don't have numbers to provide as evidence to convince your audience. Instead, your evidence is made up of pertinent quotes or observation details that illustrate your findings. You need to use these effectively and often to convince your readers that your interpretations of the data—your findings or insights—are valid. Check out Appendix B for a sample academic report in APA format.

DOI: 10.4324/9781003181750-9

Abstract

An abstract is a brief (typically 100–300 words) summary of your research. The abstract should include what your project is about, what research methods you used, and what the main findings are. Even though an abstract is presented at the beginning of a research paper, it is typically written last, after you've written the rest of the sections and can easily summarize them.

Introduction

The introduction of a research paper should introduce the reader to the topic and explain the importance of the study. Frequently, researchers use anecdotes or examples from popular media as "hooks" to draw the audience into the "story" of the research.

Literature Review

A literature review is a summary of what research has been previously conducted on the topic, as well as a brief foray into theory, the assumptions about "how things work" that guide a research project. The literature review is a substantial section of your paper and should include citations of many academic journal articles and books from the field of study. The Research Question should be presented at the end of this section and should arise naturally out of gaps in the literature review. You need to tell that reader what you plan to examine that others have not looked at before.

Methods

In this section, you outline how you conducted the study. This includes:

- Your specific method of data collection (e.g., interviews, card sorting, usability testing)
- Participant information, including:
    - Sample size, that is, the number of people who took part in your study
    - Aggregate demographics of your participants, for example, 30 male, 20 female, 40 non-binary, 10 trans
    - Selection criteria for your sample, for example, only Mac users were eligible to take part in the study
- The dates (reported as months/year) when data were collected, especially when talking about digital technologies where interfaces and features change often and quickly
- Your data analysis method—this is typically a type of thematic analysis for qualitative research

- Description of equipment or other materials used during the study, but only that which actually has a bearing on results (it doesn't matter if the participants sat on wooden or plastic chairs if you are testing a website design)

Findings/Results

This section is the bulk of your report. It is where you present your key insights, alongside the evidence for these findings (quotes and descriptions from your interviews, observations, etc.). It's important to balance description and analysis here. Explain what was said or what you saw and then present a brief analysis of "what does this mean," connected to your Research Question. Remember to keep your participants anonymous! Don't use real names or too many identifying details that could be used to link the words to the actual people who said them.

Discussion

In the discussion section, the researcher puts the findings in context and provides an in-depth analysis and interpretation of the findings, ultimately answering the Research Question. Oftentimes, the findings/results and discussion sections are combined in qualitative research reports, as the "objective" findings (what was said and observed) make more sense as a story when they are woven together with the researcher's interpretations, instead of separated out somewhat artificially in separate findings and discussion sections.

Conclusion

In this section, you present a summary of what you found and what it means. Here you can also include limitations of the current study (e.g., not having a very diverse sample) or ideas for future studies that were out of scope of this particular research.

Works Cited

The works cited or bibliography is an alphabetical list of every scholar or expert whose published ideas, thoughts, or quotes you have used throughout your report (most of these can be found in your literature review). This section should follow one of the common citation formats, such as APA (American Psychological Association) or Chicago style. For more on citation formats and style guides, see Purdue Owl's guide here: https://owl.purdue.edu/owl/purdue_owl.html

Appendices

This is where you can include any research materials that you used in the study, such as a list of interview questions.

Industry Reports

When you're working for a company or organization on the UX team, you need to provide reports of your research that have very different purposes to academic research papers. Typically, a UX researcher will provide some recommendations based on their research findings, in the form of a research summary, for the design team to use to help create or improve the app or website. Industry reports tend to be much shorter and to-the-point than academic research papers—these papers do not aim to contribute to a body of knowledge but are instead used to apply the findings internally for product development. (Sometimes industry research ends up as part of a white paper, which is a longer, in-depth document that is shared externally, that is, outside of the company. White papers present information on a particular topic while also typically advocating for a specific solution or position—usually one that benefits the company. White papers are out of the scope of this book, but we encourage you to read up on them if you're interested!)

An industry report or research summary is typically made up of a few key sections as outlined in the following, though the specifics will vary based on the needs of the organization and the specific audience. Check out Appendix C for a sample industry report.

Background

This is similar to a literature review in an academic report and includes a brief summary of any research that was done previously in your company that is relevant to your current project, as well as any other background information pertaining to the project aims and research design decisions.

Stakeholders

This typically refers to internal (to the company) people. In this section, you list all the people in your company, by name, who are part of or have a stake in the project. Some are people that need to be consulted as subject matter experts, others are part of the ideation workshops that will use these findings (these are typically UX designers and product managers), yet others are the CEO and other chiefs, who are concerned with the business goals, bottom line, and how your research project impacts both. This section can also include external stakeholders who will be affected by this research project, such as "end-users" (of course!) or external regulatory bodies that need to be consulted.

Methodology

As in an academic report, here you explain how you collected and analyzed your data, listing the specific methods used, as well as the time frame for the research.

Participants

Here, as in an academic report, you tell the reader who took part in the study, including the specific criteria for participation and the participant demographics.

Key Findings

In this section, one of the most important in the report, you write out the key findings from your study that are related to your Problem Statement. What did you see and hear as you conducted your research? Unlike academic reports, this section can use bullet points, especially in environments that are fast-paced and include a lot of iterative research for multiple projects happening simultaneously.

Known Limitations

Here, list anything that makes you less confident about your findings. This could be something like a small or not diverse sample size or limited observation opportunities.

Recommendations

This is the part where the academic report and the industry report differ. Once you have written out your key findings or insights based on the data collected, you need to make some recommendations for the design (or re-design) and development of the digital product based on these findings.

References

This is a list of all the documents (both internal and external) that were used in the production of this report.

Appendices

In this section, you provide the materials used in the study, such as interview questions, instructions that you provided to the participants, and any other relevant information that could help the reader get more detail on how the study was practically conducted.

Presenting

Often, you will be asked to present the findings of your research orally. Even more so than with a written report, you need to consider your audience. You only get one chance to present your findings—if an audience member doesn't understand, they can't re-read your presentation again! (Though, they do have the chance to ask you questions for clarification, which you should be prepared to answer on the spot.)

When Presenting Orally

First, you want to be sure you can speak confidently about your project. You are the expert in the room, and the audience is looking to you to learn about your exciting findings. The best way to ensure you can speak confidently is to be well prepared. You should create your presentation a few days in advance and give yourself time to practice what you will say. This also helps with being sure you keep to the time allotted (time yourself during your practice runs). In academia, conference presentations are often only about 10–12 minutes. This may seem like a long time, but it is common for presenters to talk more than they realize and go over time. Being well prepared means that you include everything you want to include and still leave time for questions from the audience!

Second, and speaking of the audience, be ready to answer questions! In both academic and industry settings, it is common for time to be left aside after presentations for Q&A (questions and answers). Sometimes, it can be helpful to include extra slides or content for the Q&A period that may not have fit into your official presentation time but could be helpful in the case that certain questions arise. It is important to remember that, as you answer questions, you are still in the context of your project. Stick to answers that are based in your data; don't switch to your opinion or anecdotes.

Third, it is important that you display a passion for your project. You want your audience to be just as excited about your research as you are. In academia, this can lead to more attention in the field and support for your line of research. In the industry, an obvious passion may be the tipping point for garnering the support needed to implement your recommendations!

Displaying

There are many tools or programs to create visual presentations that you can show while speaking about your project. We list some of our favorites here:

- PowerPoint—still a popular option in academia. It is part of the Microsoft Office suite, which you most likely have access to through your college or university. It provides "slides" that can be embedded with text, images, sounds, and videos. You progress through a slideshow as you walk the audience through your project.
- Google Slides—free with a Google or Google Drive account. Google Slides is very similar to PowerPoint, though it has fewer features.
- Canva—a free tool at Canva.com. Canva allows you to create slides, similar to PowerPoint, but it also provides dynamic templates and suggests images, colors, fonts, and shapes that match your presentation as you are building it.

- Mural—a free (for students) digital whiteboard. Mural is great for presenting virtually in workshop scenarios, where the team can move "stickies" around to categorize the findings, make suggestions, and answer questions.
- Zoom, WebEx, Google Teams, Skype, etc.—video conferencing apps that are also presentation tools. They allow researchers to present virtually and, with screen-sharing functionality, allow for anything on your screen to be shared. This may include a PowerPoint, Google Slides, Canva, or Mural presentation.

There are a few rules to keep in mind when creating research presentations. First, you'll want to use as few words as possible. Stick to keywords and important phrases. The goal is to get the audience to pay attention to you, not to have to read paragraphs of text in a presentation. The text is really there to act as guideposts for both you and your audience as you walk them through the project.

Instead of filling your presentation with words, use visual elements like graphs, charts, tables, and images to summarize your findings. The best presentation is embedded with interesting findings through dynamic images that help you, as the speaker, tell your story. When an audience can *see* instead of just *hear* your findings, they are much more likely to immerse themselves in, and get excited about, your research. For example, if you used the method of screenshot diaries (see Chapter 10), the presentation should contain a few, exemplar entries. These don't need to have text accompanying them; you can walk the audience through what you found, using the entries as illustrations. If you are working with text, say, for example, because you conducted a brain writing session (see Chapter 16), it is OK to paste words in as cropped screenshots to show the text from your participants in-context. This is much more exciting than just retyping their words onto your slides.

You should certainly also include relevant pictures for interest. But, be careful to not just drop in lots of random Google image search images and clipart. These can not only make the presentation sloppy but can lead audience members to wonder how this random imagery is connected to your research. This is a very easy mistake to make in tools like PowerPoint and Canva, because the programs include so many options for shapes and pictures. Remember that every piece of a presentation should be working to help the audience understand, and become invested in, your research—so choose the right images to illustrate your story.

Be sure to use colors and other aesthetic elements well. Don't use colors that clash or that may be difficult to read because of poor contrast. In academia, presentations will often be "on brand" by using a college or university's colors and logo on all the slides. In fact, many schools have PowerPoint templates for students and faculty use. In the industry, staying "on brand" may mean using the company's branding materials (logo, colors, fonts, etc.) during the presentation. Also, be careful of using anything that flashes or moves quickly, as those types of images can trigger seizures in people with certain medical conditions.

A Final Note on Differences in Presenting Academically Versus Presenting in Industry

In academia, you are considered the authority on your research and the interpretations and conclusions you draw are your own. There is less at stake in drawing the wrong conclusions in academic research than there is in industry research, where your findings inform the creation of a real product that real people use on a day-to-day basis (and impact the company's bottom line). Accordingly, in industry, presenting your findings oftentimes is a co-creation exercise—you present your raw findings through a whiteboard program like Mural and then talk through with the team what the interpretations should be. During these workshops, the team typically does some form of affinity mapping, that is, categorizing raw data into themes by moving (digital) sticky notes around. Also, remember that industry research tends to be iterative, meaning that you will do some research, present initial findings, inform the preliminary design, and go back and do more research—rarely will you have a perfectly polished, stand-alone research report that will be the final presentation of your research.

Now that you know all about Design Thinking, planning a UX research project, and how you are expected to present your findings, let's jump into learning about specific UX research methods in Section Three!

Section Three

Methods

8 Traditional Research Methods Overview

Interviews, Focus Groups,
Qualitative Surveys,
Observations, and Thematic
Analyses

The two most commonly used qualitative methods in Media and Communication Studies are interviews and observations. Many of the specific processes, tools, and skill sets that researchers draw on when interviewing and observing form the basis of the contemporary UX methods that we outline in this book. Two other methods, focus groups and qualitative surveys, are also used frequently in qualitative academic research. Qualitative data is typically analyzed using a process called thematic analysis. This chapter provides a brief overview of interviews, focus groups, qualitative surveys, observations, and thematic analysis, as subsequent chapters assume reader familiarity with these basics of qualitative research.

Interviews

Interviews are, at the core, directed conversations, aimed at finding out some information. Interviewers ask interviewees questions to better understand their thinking, feelings, experiences, and motivations for behavior.

In academia, interviews are one of the most common methods used for qualitative research in social science fields such as Psychology, Sociology, Anthropology, Communication Studies, and Media Studies. They are used in academic settings where researchers are curious about how individuals perceive their social realities. As you probably know, interviews are not just a research method—they are common in other settings such as job seeking and news articles. The difference between interviews for research purposes and interviews for jobs or journalistic interviews is that researchers typically ask *multiple* people similar questions to draw a more generalized conclusion about a particular phenomenon, rather than diving deep into only one person's story.

In UX research, interviews are usually used at the early stages of product design, in the Empathize stage, to find out what users want and need, and also to find out what they currently don't like about their experience. Because there are always other stakeholders involved in the creation of a website or app (e.g., the company that needs to make a profit, engineers that need to build

DOI: 10.4324/9781003181750-11

the product), interviews are also used in industry to understand these various stakeholders' requirements of the product.

Before an interview, a researcher needs to create an interview guide, which is a set of questions that will be asked during the interview. Typically, interviews consist of open-ended questions, such as "how" or "why" questions, to prompt the interviewee to provide in-depth, detailed answers. Interviews can be structured, semi-structured, or unstructured, depending on the needs of the research (whether the research is more exploratory or whether the researcher has a sense of the important questions they need to be answered already). In structured interviews, researchers strictly follow the interview guide—all participants are asked the same set of questions in the same order. Semi-structured interviews take a looser approach, with the interviewer adding or removing questions as the conversation unfolds. And finally, an unstructured interview flows like a conversation, with the interviewer typically coming into the interview only with some general themes they would like to cover.

It's important for the interviewer to build rapport with the person being interviewed, so that they feel comfortable sharing their insights. To build rapport, researchers might want to ask some easier questions early on, questions that are not connected to the research itself (e.g., talk about the weather, pets, and current events). During the interview, the interviewer's job is primarily to listen and guide the interaction by prompting the interviewee and asking follow-up questions. Don't insert yourself into the interview by telling too many of your own stories and sharing your own experiences! Treat the interviewee as the expert.

Interviews can take place in person, but remote interviews have been popular with researchers long before the COVID-19 pandemic forced most in-person interactions online. Remote interviews can be conducted over the phone, but video interviews (such as through Zoom or Skype) provide the additional advantage of showing nonverbal communication (such as facial expressions and gestures), which can be a source of important insights. Virtual interviews are also easily recorded, whether through the software being used or via voice recorders on the researcher's phone.

Interviews provide data in the form of transcripts (recorded interviews are transcribed into text files), which are then analyzed by the researcher, who will look through the transcripts for patterns and themes. Because interviews tend to be at least 30 minutes long, they generate *a lot* of data—a 30-minute interview creates about 10 pages of single-spaced text.

Other Ways of Asking Questions

Focus Groups

Focus groups are somewhat like group "interviews," with a researcher talking with five to ten participants at once. It's important to clarify, though, that focus groups are *not* interviews. Focus groups are more like a group discussion and are particularly valuable because of the interactions between the participants, rather than the insights gleaned from the individual answers from each participant. A small group setting can inspire participants to share personal stories and bounce off others' answers in ways that they would not have been comfortable doing in a one-on-one interview. Focus groups are useful for bringing to light ideas that might have been taken for granted, things that some individual participants might not have thought of before, but that emerge in a group setting. During a focus group, it's important to pay attention to group dynamics—the amount of agreement or disagreement around a particular topic can highlight what's important to take into consideration during product development in UX.

Typically, a research study will include at least three focus groups. (Five to eight are ideal!) Because the format of a focus group is discussion-based, where respondents are encouraged to interact with each other, the researcher serves as a facilitator or moderator, rather than as an interviewer. Focus group participants can be recruited so that groups are homogenous (where participants all share demographic and other characteristics relevant to the study) or heterogenous (diverse participants make up each focus group). This decision depends on if the researcher wants to see how different people approach, discuss, and negotiate on the same topic or if the purpose of the research is to understand how a very specific group of people understands or feels about a topic.

A focus group moderator will use a focus group guide, a pre-prepared list of prompts, questions, and activities, to help facilitate participant interaction, during the discussion. As with interview guides, the first questions should be broad and easy, so that participants can feel at ease in the group and to encourage sharing. Focus group guides are particularly important when facilitators other than the primary researcher are used across different groups in the same study. This ensures that the questions asked and prompts used are the same in all the groups, to provide comparable data for analysis.

Focus group moderating can be a difficult task. You need to guide the conversation in a useful direction related to the Research Question or Problem Statement at hand, which can be hard to do with many voices at the table, each with their own stories and opinions. The facilitator also has to make sure that one particularly chatty participant doesn't dominate the conversation and that shy participants are encouraged to speak up. This is why a focus group moderator should be a skilled communicator and a credible presence to the focus group, so that they can (gently but firmly) assert their authority and keep the conversation on track. In UX, focus groups are used to assess user needs and feelings about a particular product. In academic Media and Communication Studies research, focus groups are really helpful for understanding group norms and dynamics around communication.

Qualitative Surveys

Surveys are another way of gathering attitudinal information from participants. Surveys consist of questions and scales (sets of questions that measure a particular attribute, such as a personality characteristic or digital literacy) that get at the who, what, and where of your users. In academic research, surveys are usually considered a quantitative method, in that they gather information that can be analyzed and presented numerically, such as demographics, numbers around media or technology use (time, frequency, etc.), and scales where respondents can rate an option on a scale from 1 to 5 (strongly disagree to strongly agree). Qualitative surveys, on the other hand, ask open-ended questions of participants (typically asking how or why), similar to interviews, but in this case, participants respond in written format, without an interviewer present. These types of questions demand long-form written answers, ideally paragraphs, that tell an anecdote or shed light on participants' feelings and attitudes around a particular topic.

Surveys are usually conducted online using programs such as SurveyMonkey or Google Forms, where a link is sent to all participants to answer the survey questions at their convenience. Surveys can also be sent out via mail (with a stamped return envelope for participants to send back the completed questionnaire) or conducted over the phone (typically using automatic systems). No doubt you've experienced a survey before—for instance, after calling customer service, it is common for a quick two- or three-question survey to be presented to callers, asking them to rate their satisfaction with the interaction, from 1 to 5.

Surveys are used in UX research to solicit feedback from users, either in the Empathize stage (to understand users' current needs) or in the Test stage, where researchers want to evaluate a website or app. Surveys are often built into the digital product itself—through a "popover" with a couple of questions that pops up while a person is interacting with the website or app. When conducting qualitative surveys, you should keep the questions minimal—one or two questions asking for written answers are the maximum that you should ask for users to answer, particularly if they're presented in a popover format (e.g., tell us what we could improve about this website . . .). Too many questions can turn off the user—how often have you started a survey and then stopped because the question set was dragging on and it seemed like too much effort?

The main limitation of using interviews, focus groups, and qualitative surveys for UX research is that, while they give you great insights into how and what people *think* (attitudinal data) as well as self-report data on their behaviors and habits, they do not give researchers insight into how people actually *behave* (behavioral data). Often, what people do and what people *say* they do are different things. When you're concerned with how people use a digital product, observations can provide more accurate data of *actual* use.

Observations

Observations, as the name suggests, involve a researcher watching and listening to people. Observations have historically been conducted in person, but as more and more of our lives have gone digital, researchers do more online observations or digital ethnographies, observing people in digital spaces.

Observations are a very common technique in UX research because ultimately UX researchers are interested in understanding how people interact with something—how they *actually* behave—so that that experience can be improved. In UX, observational research takes place during the Empathize stage, when researchers want to discover how people currently use a particular product, but also during the Test stage, when researchers want to evaluate whether a product is easy and efficient to use. As you'll see later in this book, many different UX methods have an observation component.

In academic studies, especially in Communication and Media Studies, observations are used for a similar purpose, to discover and explain people's behavior, particularly regarding communication devices and consuming media. As mentioned previously in this book, the key difference between academic research and UX industry research is that academics are not typically working to provide recommendations for the creation of a new media or communication product or product improvements but instead are adding to the body of knowledge about when, how, or why certain media is used or certain communication patterns occur. For example, observations as part of an academic research study can show how different family members may use a plethora of devices for entertainment all in the same space. This sheds light on changing cultural trends, such as the shift from a family sitting down to watch the same TV show at dinner (common before the rise of mobile technology devices) to each family member choosing their own entertainment on a different device and being together in the same space but not consuming the same content.

Observations as a method in both academic and UX research range from quick observations of people completing specific tasks to months-long or even years-long ethnographic field studies. Short observations can be done in the lab or in the field (in the place where people would typically use the product in real life). These types of observations are usually concerned with a better understanding of how a person uses a specific device or product. In UX, the most commonly used limited time observation method is usability testing, where a researcher observes a user navigating through a website or an app and takes note of pain points or issues.

Longer observations typically take place in the environment of the user, where a researcher lives and/or works with the people they are studying, observing every move and trying to understand the culture as a whole. For example, a UX researcher might do an ethnographic study at a college, observing how students go about their daily lives on campus, including going to lectures and socializing, to inform the design of a new university website. Observing students on campus can provide many rich insights that can lead to ultimately creating a better website. For example, perhaps the researcher sees that students often get lost when trying to find lecture halls. Based on this observation, they might recommend that a clickable campus map, with clearly labeled classrooms, is added to the home page of the university student portal, so that students have that resource handy as they're navigating the physical environment of campus.

Observations can be messy with the huge amounts of different types of data collected, particularly during an ethnographic study. A solid ethnographic study will yield hundreds of pages of notes, photographs, videos, recordings, artifacts (objects or documents from the culture or place being studied), and more. The scope of the observation will dictate what data are collected. For example, a usability testing session where the researcher watches a user navigate to a specific page on a website will be more manageable to document (simply record the screen and note pain points) than an ethnographic study of a college campus, as described earlier. This is why researchers need to have a solid plan prior to conducting the observation, so that they know what to focus on while they're in the field.

Field notes are detailed notes that an observer takes while they are observing, jotting down details such as who is being observed, their behaviors and interactions, the physical environment itself, the participants' movements within the space, etc. Recording observations, whether via video or audio, as much as possible is also helpful. Ultimately, you want to focus your observations on what you need in order to answer your research questions or help fulfill your research objectives. This might seem obvious, but it's crucial to carefully think through what sort of data you might want to hone in on. For instance, if you're wanting to understand interpersonal communication patterns between students at a coffee shop, you will want to pay attention to interactions between groups of people more so than observing solo students go about their business.

Often, observations are conducted in conjunction with interviews (read more about contextual inquiry, a type of UX method incorporating observations with interviews, in Chapter 12). Interviews conducted before, during, or after observations can provide researchers with insights into the motivations for and thought processes around user behavior.

Analyzing Qualitative Data: Thematic Analysis

Qualitative data use words, not numbers, and so analyzing qualitative data is more time-consuming and messier than simply running statistical tests on survey answers. An established way of researchers analyzing qualitative data is through thematic analysis. Thematic analysis is what it sounds like: organizing the collected data by themes that emerge across the different interviews, focus groups, surveys, or observations. A similar process of grouping gathered data into themes and categories is called "affinity diagramming" or "affinity mapping" in UX.

Thematic analysis uses three criteria for finding patterns and themes in vast amounts of qualitative data: recurrence, repetition, and forcefulness. As you're reading through pages of interview or focus group transcripts, question answers in qualitative surveys, or field notes from observations, you should pay attention to things that have a similar meaning across participants (recurrence)—for example, multiple participants talking about similar topics even though they are not using the same words; the same words being used repeatedly (repetition); and forcefulness (listening for volume and inflection changes, as well as pauses in the conversation). Paying attention to these three criteria can help you pull out what is salient to participants in your research.

You can conduct a thematic analysis in a Word document, by using comments and color to group findings and categorize themes. There are also specific software programs for qualitative data analysis, such as NVivo or Atlas.ti—but it is much more expensive than simply using Word. You can also conduct analyses of qualitative data using paper/sticky notes and physically grouping the notes, or you can use a digital whiteboard, such as Mural. Mural is often used in UX industry research for affinity mapping, where digital stickies are moved around and organized by theme.

Final Thoughts

As you read through the UX methods in this book, you will notice that we often mention asking "good" questions of your users or paying attention to participants' behaviors. All of these integral pieces of more contemporary UX methods are still rooted in traditional qualitative research methods discussed in this chapter. Critical thinking has always been necessary in research, and that is no different for UX. The mindset necessary and the methods used may seem very different to academic research, but really UX methods are, at the core, innovative versions of asking questions and observing how people act. And, just as in traditional academic research, UX data analysis includes time spent reviewing and re-reviewing data, finding patterns and themes in the data, and using your expertise and previous studies' findings to arrive at your insights.

Further Reading

"How to Conduct User Observations," *Interaction Design Foundation*. www.interaction-design.org/literature/article/how-to-conduct-user-observations.

Brennen, Bonnie S. *Qualitative Research Methods for Media Studies*. New York: Routledge, 2017.

Lindlof, Thomas R., and Bryan C. Taylor. *Qualitative Research for Communication Methods*. Los Angeles: Sage Publications, 2017.

Pernice, Kara. "User Interviews: How, When, and Why to Conduct them," October 7, 2018. www.nngroup.com/articles/user-interviews/#:~:text=A%20user%20interview%20is%20a,of%20learning%20about%20that%20topic.

9 Emotional Journey Mapping

While having an app or a website function properly is obviously important, there is also extreme value in a space providing an enjoyable and rewarding *experience* for users. After all, we are talking about user experience! A really great way to understand how a space makes users feel is to map their emotional journeys. During emotional journey mapping, you, as the researcher, follow the "path," the sequence of events, or the "touchpoints" that users move through while using the space. Importantly, with each step you note how pleased or frustrated they are. Mapping experiences to a chronological chart helps to visualize what the app or website is doing right—the "highs"—and what isn't going so well—the "lows."

Quick Tips	
Tools	Whiteboard and Post-It notes or spreadsheet software like Excel
Use When	You want to understand how a space makes a user feel; focus on emotions and improving experience. Your research project involves emotions around technology use, media consumption, or communication processes (for instance, you can map the emotional journey of interpersonal communication during a first date!).
Design Thinking Stage	Empathize

The Method

Emotional journey mapping invites participants to take the researcher on a sort of tour. Also known as customer journey mapping, emotional journey mapping is primarily concerned with understanding a user's complete experience from opening an app or a website to closing the app or browser tab.

The first mention of the idea of mapping a customer's journey is in Chip Bell and Ron Zemke's 1989 book, *Service Wisdom*. Calling it the "cycle of service

DOI: 10.4324/9781003181750-12

mapping," customer journey research began by encouraging researchers to map experiences in physical spaces.[1] A popular application was the 1999 study of what would eventually become the Acela high-speed train in the northeastern US. IDEO and C+CO conducted emotional journey mapping to capture a riders' complete experiences with the service.[2]

To conduct an emotional journey map, *touchpoints*, or each area visited or task completed, are mapped chronologically. For example, a customer journey map of a grocery store may include parking, store appearance, entry, produce section, deli, bakery, canned goods, and checkout. At each of these touchpoints, researchers note how pleasing or frustrating the experience was, how long the participant stayed in that space, what they did while there, and any responses to interview questions. After mapping several journeys, researchers look for trends: What is the grocery store doing well and where are shoppers having an enjoyable experience? What can the store improve and where are shoppers getting frustrated? Where are the highs and lows?

Maybe, after mapping several journeys, researchers notice that most participants are fine moving through the store, but checkout is frustrating. This may cause them to rate the grocery store poorly overall because checkout is the last thing they remember. But, researchers can now suggest that the store look into how they can fix the checkout experience. Then, the overall ratings and experiences may drastically improve.

As another example, maybe researchers notice a trend that the parking experience and the appearance of the storefront lead to annoyed shoppers. Subsequently, no parts of the shopping journey are gratifying because the first few minutes were so unenjoyable. Through this example, researchers may suggest that the store research how to improve their parking and storefront before making internal changes, to see if a better early impression doesn't help to improve the entire experience.

Of course, other trends may be found somewhere in the middle. If a task takes longer than a certain amount of time, that may mean the difference between an enjoyable experience and an annoying one. This could help the grocery store recognize where they may need to place extra staff during certain busy times. For example, shoppers may generally feel gratified at the deli, but once the wait is longer than ten minutes, they typically feel it is a bad experience. Tracking when the participants are shopping, researchers realize that this long wait is common on Sunday afternoons. The store can now staff the deli with more workers on Sundays or research other ways to make the Sunday deli shoppers happier.

Emotional journeys for website and app users work in a very similar way. As the researcher, you map how users move through the digital structures, noting what areas or tasks are enjoyable and which are frustrating. Touchpoints represent steps in a digital process like changing a password, posting a photo, scrolling through a news feed, or attempting to purchase a product. Indeed, touchpoints need not be formal procedural steps, but they can be.

The nice thing about emotional journey mapping is that it is a very flexible method. You get to decide exactly what you are measuring. First, the researcher decides what the emotional spectrum will be based on the research problem, the company's mission/goals, and the tasks being mapped. For example, the spectrum may go from entertained to bored, from satisfied to disappointed, or from pleased to frustrated. Second, the researcher chooses what types of data to collect beyond the touchpoints—times, notes, images/screenshots, and so on. Third, the researcher also decides what the journey will be. Open-ended journeys invite the participant to use an app or website as they normally would. In other words, they implicitly define what the touchpoints are as well as their order. In a close-ended journey, researchers decide on the same list of touchpoints, in the same order, for every participant. A combination of the two can also be used. Maybe you give each participant the same goal, but let them create the path to that goal on their own.

Mapping users' emotional journeys is similar to traditional academic methods like observations and ethnographies. Like observations, the goal is to watch as participants do the relevant task. As a researcher, it is your job to note as much as you can about the experience. Similarly, like ethnographies, user emotional journeys are not just about collecting objective, surface data like which touchpoint is visited, when, and for how long. Journeys aren't really even just getting a Likert-scale-like number that represents emotion. Instead, as with ethnography, as the researcher you want to understand what the experience is like for the participant, empathizing with their highs and lows and recognizing why things may be satisfying or frustrating, even if the participant isn't quite sure why.

Use When

Emotional journey mappings are typically employed at the beginning of a study, in the Empathize stage, usually before a research problem is formed. This is because emotional journey mappings are meant to uncover how a user interacts with a product, how the product makes them feel, and what specific touchpoints are enjoyable and which are frustrating. Thus, emotional journeys are best used when you are beginning your user-centric project. The method can tell you a lot about what an app or a website is doing well and what aspects are providing subpar experiences.

In academic research, emotional journey maps can be used to visualize a person's experience. For example, if you are studying interpersonal communication at a doctor's office, you can ask that your participants describe their emotions through each touchpoint—parking, entering the building, checking-in, filling out forms, waiting in the waiting room, reading any signage, going into the exam room, communicating with the nurse, talking to the doctor, being examined, checking out, and leaving. The emotional journey maps can help researchers learn about common interpersonal experiences at doctors' offices.

For instance, perhaps the study can be used to compare experiences at different offices or between different groups of patients.

What You Need

Creating journey maps does not require impressive or expensive technologies. At the most basic level, maps can be created by hand, on pieces of paper, or on whiteboards. Live-sketches from participant sessions can be cleaned up and compared in person or scanned for virtual collaboration.

A surprisingly useful digital tool for creating journey maps is Microsoft Excel. Utilizing simple functions like graphs and cell colors, you can create visually appealing and efficient maps that can also be easily blended to reveal means and trends. Begin by creating charts that represent standardized touchpoints and emotion tiers. Then turn this chart into a line graph. You could just stop there, but to make a true emotional journey map, you can use options like background gradient coloring, moving your touchpoint labels, arrows, and other tools to create a visually telling map that easily shows highs and lows and that can easily be compared between participants.

In Figure 9.1 you can see a sample emotional journey map. Let's say you are studying Canvas, a learning management website. You decide on seven touchpoints prior to interviewing participants—logging into the website, going into a course, reading the syllabus, sending a classmate a message, reading a PDF, watching a class Zoom recording, and submitting a course assignment. Although the image is black-and-white, a colored gradient to represent "gratifying" to "horrible" experiences is recommended. In this example, the top is usually green; moves into blue, yellow, and orange; and then ends in red.

We can see from the figure that this particular participant's average experience using Canvas is generally enjoyable. However, attempting to watch a recorded Zoom meeting was described as a "horrible" experience. In addition, it seems that this frustrating experience may also affect any experience that comes after—this participant described already being annoyed, having to get back into the Canvas course space, and then maybe not completely understanding how to complete or submit the assignment because they either couldn't watch the recording or were just generally annoyed. Just looking at this one map, a Problem Statement seemingly should focus on the recording viewing experience, but, of course, a few more maps are needed!

Analyzing the Results

As you collect emotional journeys from your participants, it is important to remember that they will not all look the same. Indeed, the goal isn't to obtain a lot of maps that all look exactly the same and expect the next step to be obvious. Of course, you will take note of when a touchpoint dips into the "bad" or "horrible" area. But it is also fruitful to look at what experiences are in the

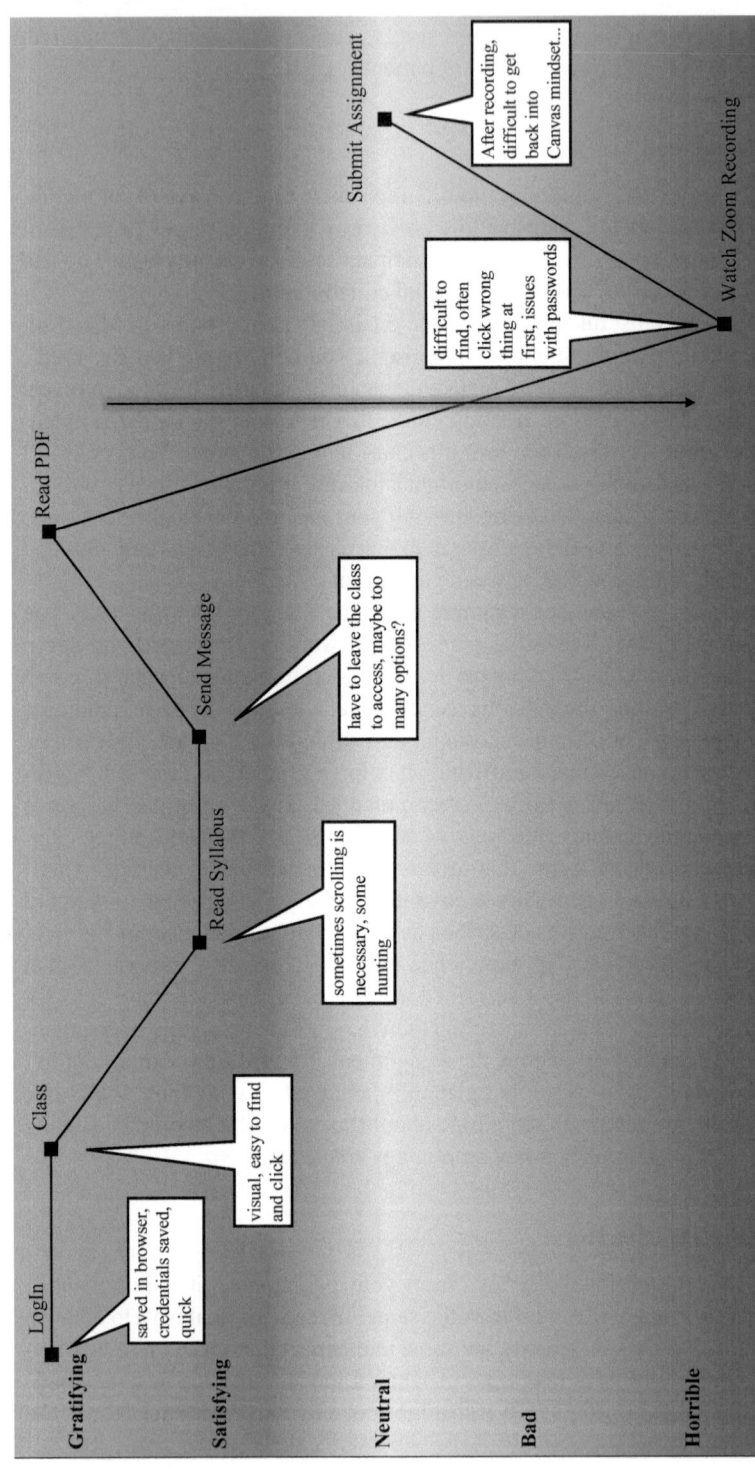

Figure 9.1 A Canvas emotional journey.

"gratifying" and "satisfying" zones and why. Learning what are pleasurable experiences for users can help in two main ways. First, you can learn methods for addressing frustrating points by knowing what users enjoy. Second, you can be proactive and test changes to tools and functionalities that aren't necessarily large issues yet but could still benefit from alterations based on what users are already enjoying about your product.

As you are looking at each map's low touchpoints, try to think about the map holistically. Consider variables like what participants said about them, how long participants generally were at that touchpoint, what they are doing before and after that touchpoint, and how important that touchpoint is to both the users and the overall experience. Again, the goal is not to find the perfect experience through emotional journey mappings but to begin to see the experience through your users' eyes, learning how they *feel* when they are within your app or website.

Case Study

In 2018 Nike used emotional journey mapping to study customers' experiences with using the company's new app and then entering a brick-and-mortar store. Michael Martin, VP of Digital Projects at Nike, has a background in entertainment and gaming products, and for Nike's new "Nike Direct" product, Martin first studied how customers experience the brick-and-mortar Nike sneaker shopping experience by mapping touchpoints and their delights and frustrations.[3]

He found that customers in a physical Nike store experience peak excitement when they find a sneaker they like. But this can quickly flip as they wonder things like "is this the only color?" and "what if they don't have my size?" Because of emotional journey mapping, Martin was able to identify and suggest solutions for these lows. He then used these findings to make recommendations for the new product—Nike Direct. On this app, customers can see a product they like, scan it, immediately know about color and size options, and request that a salesperson bring out the customer's choice fairly quickly. Customers can also reserve time slots, scheduling one-on-one time with an employee for an almost personal-shopping-like experience. Martin used journey mapping to understand when Nike customers are traditionally pleased and frustrated and helped to create a data-driven app that worked on increasing the highs and decreasing the lows.

Steps

1. Decide what you will be mapping.

    a. Is it a complete experience, from start to finish like using the app on a normal day?

    b. Is it using a specific tool or functionality like changing privacy settings or searching?

    c.   Is it implementing a specific process like signing up or posting a photo?

2.   Decide if you will lay out the touchpoints ahead of time or if you want the participants to create their own paths.

3.   Observe as the participants complete the process, noting touchpoints, time spent at each touchpoint, emotions felt, and context of usage.

    a.   If you can be physically with the person, you can watch over their shoulder in a more casual setting. Or, in a more lab-like setting, you may connect a laptop or phone to a larger screen that you can watch through screen mirroring. Ask for screenshots or take your own photos of the process if you can.

    b.   If you cannot be with the person, you can complete the mapping over a program like Zoom. The participant can either perform a laptop hug (holding their laptop backward, on their lap, so the camera picks up what is happening on their phone) or, if the study is web-based, they can share their screen. Take screenshots if you can.

4.   Ask good questions.

    a.   Be sure to include thought-provoking, but unobtrusive, questions.

    b.   Provide cues for participants who are less chatty and provide guidance for participants who may begin to include irrelevant information.

    c.   Ask participants why they made a certain choice, why they stayed at one touchpoint for a very long, or very short, period of time, and why they felt the emotion they did.

5.   Create maps of touchpoints on an x-y axis, allowing the x-axis to represent time and the y-axis to represent emotion. Attach relevant information to this map including notes and images/screenshots.

6.   Analyze the mappings by comparing variables like emotion and time spent. Remember to consider varied user-types. Try to find trends, noting where it seems a lot of participants may be unsatisfied. You should also be noting trends related to where participants feel extremely satisfied—these insights can help improve other pieces of the app or website.

Discussion Questions

1.   Think about a task you perform each day. It could be showering, brushing your teeth, driving to school, making dinner, or something else. What are the touchpoints? How do you feel at each one? Satisfied? Frustrated? How could you improve your experience of that task by altering one touchpoint?

2.   Working with a partner or group, consider the same daily task. Do you all have the same touchpoints? Does each touchpoint mean the same thing or have the same significance to everyone in the discussion group? Are frustrating touchpoints the same for everyone in the discussion group?

Notes

1. Ron Zemke and Chip R. Bell, *Service Wisdom: Creating and Maintaining the Customer Service Edge* (Minneapolis: Lakewood Books, 1989).
2. "The Story of the Journey Map—The Most Used Service Design Technique," *International Service Design Institute*, 2020, https://internationalservicedesigninstitute.com/the-story-of-the-journey-map-the-most-used-service-esign/.
3. Yasmin Gagne, "Nike's New Concept Store Feeds Its Neighbors' Hypebeast and Dad-Show Dreams," *Fast Company*, July 12, 2018, www.fastcompany.com/90201272/nikes-new-concept-store-feeds-its-neighbors-hypebeast-and-dad-shoe-dreams.

10 Screenshot Diaries

A popular traditional qualitative method is diary studies. Asking participants to keep track of, or journal, certain activities or feelings allows researchers to understand a phenomenon both over time and through a personal, contextual lens. Screenshot diaries do just this, but they also add the visual element of screenshots. Participants are asked to take screenshots of predetermined or salient moments of product usage and to then caption and annotate these images. The outcome is a look-book that dives into a user's everyday experiences with a product.

Quick Tips	
Tools	Google Drive
Use When	You want to understand users' everyday experiences in context and over some set period of time.
	Your research project involves tracking how people experience a digital communication or media phenomenon over a longer period of time.
Design Thinking Stage	Empathize

The Method

In traditional academic research, the diary method is a sort of assignment. After recruitment, researchers ask their participants to begin journaling, or keeping a diary, related to the phenomenon being studied. For example, a Media Studies researcher may want to better understand how college students consume streaming services like Netflix and Hulu. Using a diary study, they can capture topics like content viewed, when, and for how long, as well as more personal and contextual information like why that content was chosen, why that content was viewed at the specific time, how the participants felt while viewing, and so on.

DOI: 10.4324/9781003181750-13

In diary studies, entries are written by hand in a notebook or journal provided by the researcher. The participants then return their diaries, with all of their entries, once the allotted time period has ended. Today, with virtual tools, participants may be asked to keep their entries on their computer and email them at the end of the study. Or, if the researcher wants to see entries in real-time, they may ask participants to email an entry daily or weekly, or to complete the entry in some shared space like Google Drive.

Employing a diary study makes results much more personal, largely because the participants are in their typical, natural environments. The only change to participants' routines is completing a diary entry at the end of the day or after they have finished the related task. Because the participants are not in a lab or sitting face to face with an interviewer, they are more likely to contribute authentic data. In addition, diary studies are often performed over a longer period—a week, a month, a year. These longitudinal data help to understand how participants experience something over time. In a weekly study, researchers could find trends related to different days of the week. In a longer, maybe a yearly, study, researchers may find trends related to weather changes or holidays celebrated.

In traditional diary studies, the parameters for what and how a person diarizes are up to the researcher. When participants log an entry and what they include in that entry should be directly related to the study's goals, questions, or hypotheses. Some diary studies are open and simply ask participants to jot down whatever they feel is relevant to a topic. Other diary studies include specific questions to be answered at specific times during the day or after specific tasks have been completed. So, in our streaming services example, participants may be asked to complete a diary entry every night and to respond to four questions: What did you watch today? For how long did you watch it? When did you watch it? Why did you choose that content to watch? Or, they may simply be asked to generally write about their streaming experiences at the end of each week.

This traditional mode of diary studies can certainly be used in UX. You could ask participants to complete entries about their experiences with an app or a website, or you could ask them to focus on a specific tool or functionality. For example, if a research problem has to do with making a particular online shopping experience better (and more regular, from a business sense), participants could be asked to write down if they bought anything from that specific website, what it was, when exactly they bought it, and why. Answering these questions over weeks or months allows researchers to understand shopping trends.

Screenshot diaries are a new visual twist on traditional diary studies. The setup remains the same. As the researcher, you decide the parameters for the study—again, these should be related to your research problem and should consider the time and resources available to you. However, instead of just asking participants to journal, you ask the participants to take screenshots throughout the day and to then caption and annotate these images.

A great way to conduct a screenshot diary study is through Google Drive. Set up folders for each of your participants (being sure that the privacy is set so that each participant can only access their own folder). You can put the prompts and questions right in the documents, and then all your participants need to do is drag or paste in their screenshots to the folder every day. Google Docs tools make it easy to annotate and caption an image once it is in the file. As with traditional diary studies, you can ask your participants to simply take screenshots whenever something they view as important happens within an app or a website. Or, you can be more specific, asking that they take a screenshot whenever they are frustrated or to take a screenshot of the first thing they do when they visit a site or open an app. You can even ask them to take screenshots of, for example, their work computer desktop at specific times during the day. Because using Google Drive allows for more organization, you can, if it's relevant to your study, ask that each image includes a date and time.

Let's think about examples related to Instagram. As one instance, perhaps a research problem is focused on users' frustrations with targeted content. The researchers could ask the participants to take a screenshot whenever they experience frustrating content. They would normally be asked to also include the date and time, as well as a caption and annotations to explain what is going on. Figure 10.1 is an example of this type of entry.

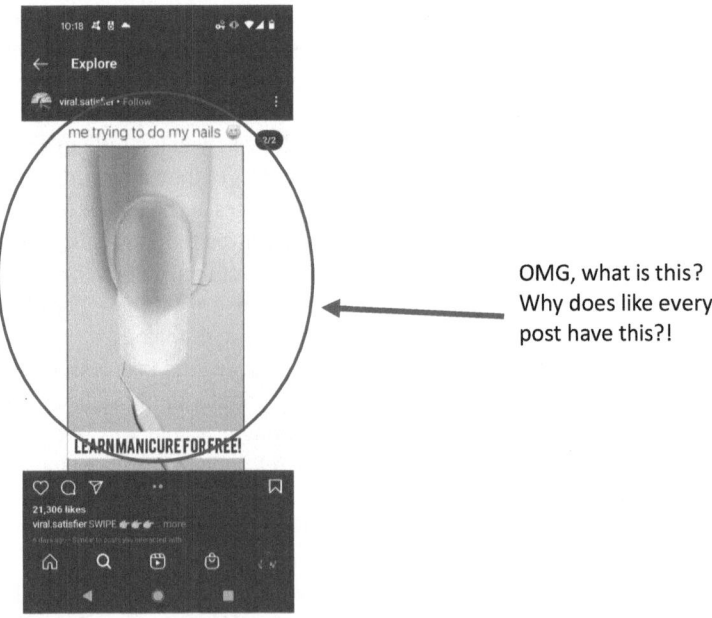

March 4, 11:35pm
Scrolling through Discover trying to fall asleep.

Figure 10.1 Screenshot diary entry showing when a user is frustrated with certain content.

As another example, imagine a research problem has to do with the functionality of updating a password. For this specific task, you could ask your participants to screenshot each step in the process of updating their password and to explain what they were trying to accomplish and how they were feeling. Figure 10.2 shows an example of a screenshot diary entry related to changing passwords in Instagram.

It is incredibly helpful to give your participants examples so that they know what you are expecting from them. (Feel free to use the two above as is or with some changes!) If you want really long captions that are more like traditional diary entries, provide examples that display that. If you want lots of annotations with emojis, arrows, and circles, provide those types of examples.

Be sure that your participants know how to take a screenshot on their device. Different mobile devices have slight variations in buttons pressed when it comes to taking screenshots. A PC laptop is slightly different from an Apple. It's probably best that participants don't literally take a shot of their screen with another camera. Those images are never clear, and it becomes more work for the participants to get the images into their Google Drive folder. A few tips before the study begins or a how-to guide in the participants' Google Drive folders can go a long way!

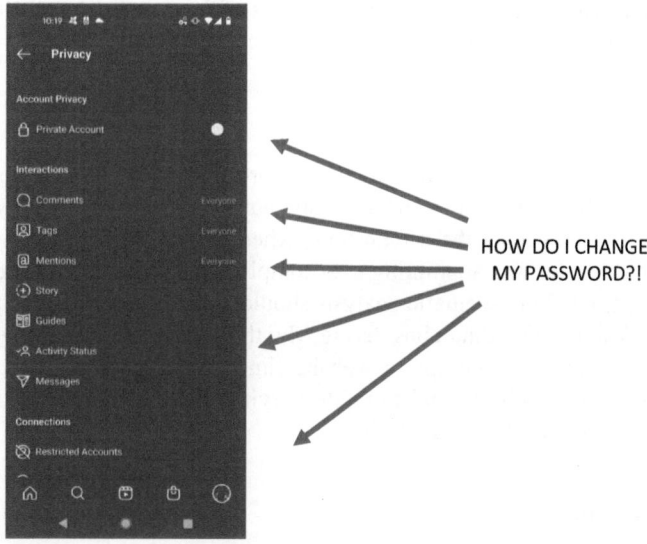

March 4, 3pm
Had some weird stuff happening with my account so thought I would update my pw

Figure 10.2 Screenshot diary entry showing that a user cannot find the password change setting.

Use When

Like other Empathize stage methods, screenshot diaries are great to use when beginning a new study, when you're trying to figure out users' relationships with your product. Screenshot diaries give personal and authentic looks into users' everyday lives—how they use your website or app and the way that your product makes them feel as they move through their daily routines. Screenshot diaries are one of the few UX research methods that send participants off on their own for the entirety of the study, so participants have very little contact with you, the researcher. This means that they are less likely than in other studies to feel like they are being studied and more likely to share their true feelings, providing you with a very human-centered perspective.

What You Need

Screenshot diaries work really well as a virtual method. Because it already involves digital images and participants contributing during their everyday lives, it does not require any in-person meetings. You will need some way for participants to organize and submit their entries. We suggest using Google Drive. Creating a Google account is free, and many universities and companies offer official Google Drive accounts to employees. It works best to create a folder for the study, and to then create a folder for each participant. Then, share each participant's folder with them. You will also need some instructions for the participants as well as some examples of the types of screenshots, annotations, and captions you are hoping to collect.

Analyzing the Results

Analyzing screenshot diaries includes more than just looking at the images that participants have submitted. A holistic analysis means thoroughly examining what is included in the screenshots, when the screenshots were taken, what the participant was attempting to accomplish, and the screenshot captions and annotations. Your thematic analysis should consider all pieces of participants' submissions, including considering who those people are and how they seemingly experience your app or website similarly and differently. Remember to consider accessibility and possible varying cultural understandings of your product as part of your analysis.

Case Study

Researchers from the University of California, San Diego, and the Nokia Research Center partnered to better understand how mobile device users experience digital history across their personal devices. Even though cloud services allow data to be saved for later access on any device, previous research had found that users still re-search and re-access content anew when they move to a different device. To study how users re-access content across devices, the researchers used screenshot diaries.[1]

Fifteen participants were asked to take screenshots when they re-accessed content on their mobile device. They were also asked to annotate them later in the day, through a nightly journal, for which participants were sent daily reminders to complete. The researchers found interesting patterns related to where content was first accessed and where it was re-accessed. For example, when content is first accessed on a computer and then re-accessed on a mobile device, it is likely because users are showing their friends that content. On the other hand, when content is first accessed on a mobile device and then re-accessed on a computer, it is likely that users experienced some technical barrier that forced them to move to a device with more capabilities. Overall, planning ahead to efficiently share content across devices is a lot of work and is not often considered at the time, leading users to then waste time re-finding and re-accessing content.

Through their findings, the researchers propose three main solutions. First, content could be identified as likely to be re-accessed and tagged with location and time content to then reshow that content to the user at a better time. Second, tools should allow for better automatic sharing of bookmarks and web history. Third, content should be freed from "application silos."[2] For example, if someone looks up directions via Google Maps, that content should automatically be saved to the user's mobile device.

Steps

1. Set up a Google Drive folder that is exclusively for your screenshot diary study.
2. Decide what you want from participants. How many entries and how often? Will entries be more open, or do you want to list out specific things your participants should take screenshots of?
3. Create a study instructions document, as well as a how-to guide including sample entries and steps for taking a screenshot on multiple types of devices.
4. Create folders for each participant. Each folder should contain the documents created in Step 3.
5. Contact participants, linking them to their personal folder and directing them to the instructions and how-to guide.
6. Once submissions are complete, code each entry by participant and number. Then, you can easily move them all to one folder so they are more easily read but not mixed up.
7. Look for trends related to what participants took screenshots of or how they felt about the specific screenshots you asked them to take.

Discussion Questions

1. How are screenshot diaries similar to emotional journey mappings? How are they different? Come up with a scenario where you would use a screenshot diary as well as a similar scenario where mapping an emotional journey would make more sense.
2. Screenshot diaries are usually considered an Empathize method. What steps can you take to ensure that participants are revealing their emotions related to experiences within your app or website during a screenshot diary study?
3. Think about an example Problem Statement that would warrant a week-long screenshot diary study. Now think about an example Problem Statement that would warrant a six-month-long screenshot diary study. Discuss how these two Problem Statements are different with a small group.

Notes

1. Elizabeth Bales, Timothy Sohn, and Vidya Setlur, "Planning, Apps, and the High-End SmartPhone: Exploring the Landscape of Modern Cross-Device Reaccess," in *International Conference on Pervasive Computing*, eds. Kent Lyons, Jeffrey Hightower, and Elaine M. Huang (Berlin: Springer-Verlag, 2011).
2. Ibid., 16.

11 Breakup and Love Letters

It is no secret that users have a love/hate relationship with technology. In particular, many users are bound to their favorite digital products, such as smartphones and apps, almost as if these devices were a partner or spouse! People depend on these technologies, expecting them to be there when they need them. Asking participants to write a breakup letter or a love letter fleshes out these relationships and reveals what users truly love, and hate, about your product.

Quick Tips	
Tools	Paper and pens; Google Docs and Zoom
Use When	You want to understand why users are loyal to your product or why they may leave. Your research relates to understanding people's (perhaps subconscious) emotions—and the reasons for those emotions—around a particular technology.
Design Thinking Stage	Empathize

The Method

As we discussed with the Heinz ketchup study in Chapter 1, just asking participants to talk about how they use something through an interview or a survey rarely reveals actual day-to-day experiences, delights, and frustrations. Often, directly asking your study participants to list what they love and what they hate only gets them talking about surface-level tools and functionalities that they think you want them to mention or that they may have read about in recent news. Instead, asking your participants to write either a breakup letter or a love letter taps into personal feelings that begin to reveal what really keeps users invested and what may make them sign off for good. Often, participants relive salient moments that would have gone unmentioned through more traditional methods like interviewing.

DOI: 10.4324/9781003181750-14

In 2009 Smart Design, a strategic design company, came up with the idea of having research participants write letters either breaking up with or showing their love for products and brands. The goal was to create a method that better tapped into the relationship that users have with products and brands. They found that asking participants to write letters, addressed to products instead of people, helped users communicate their loyalties and frustrations in ways that interviewing couldn't. In 2010 they piloted this idea, asking some conference participants to write breakup letters.[1] You can watch some clips here: https://vimeo.com/11854531!

Asking participants to write a breakup or love letter usually works best when you ask them to address the letter to a product or broad experience instead of one tool or functionality. Therefore, it is best used at the beginning of the design process, most often during the Empathize stage, or as an exploratory method for a research question that focuses on participant experiences or emotions. For example, you could simply ask participants to break up with Instagram or write a love letter to a video game. Or, if you are interested in studying a brand, participants could be asked to break up with Google or Apple. If, as a Media Studies researcher you are interested in, say, viewers' experiences while binging a show on Netflix, you could ask that they write a love letter or breakup letter that speaks to their experience.

Once you decide on the topic for the breakup or love letter, participants should be given about ten minutes to complete their letter. Before they begin, it is often a good idea to provide them with an unrelated example—if you are studying an app, show them a letter addressed to a food or clothing brand. This way your participants will know the type of letter you are expecting them to write but they won't be swayed to use some of your ideas. Usually the method includes deciding if all participants will write a breakup or love letter. Or, you may decide that half of your participants will do one and the other half, the other. Another method involves you letting participants decide which type of letter they would like to write. In any case, best practices include only asking your participants to write one or the other. Asking each person to write both a breakup letter and a love letter doesn't allow them to focus and immerse themselves in the emotions involved.

Writing letters can be completed in-person or virtually. If completed in-person, participants can choose how they would like to write the letter—in their own notebook, on a piece of paper you provide, on their laptop, etc. If completed virtually, participants can use tools like Google docs or email to get their letters to you. In both cases, you want to collect the letters for analysis.

If possible, an incredibly powerful second step is to ask participants to read their letters aloud. Being able to observe facial expressions, body language, and tone adds even more insight into what the participants are thinking and feeling. This second step is not always possible but if there is any way to listen and watch your participants read their letters, make it a part of your study.

Writing letters together can act as a sort of focus group. Five to ten participants can be in the room together, silently write their breakup or love letters for about ten minutes, and then, one by one, stand up and read their letters. If holding the study virtually, you can conduct it synchronously, meeting with five to ten participants over a program like Zoom. Letters can still be written silently together, and then participants can take turns reading what they wrote. If synchronous focus groups are not possible, you can have participants read their letters aloud to just you. Or, if meeting synchronously can't happen at all, participants can record themselves reading their letters.

Use When

Breakup and love letters are best used in the early stages of UX research, typically in the Empathize phase, to tap into salient experiences that speak to why your users are loyal or why they may leave. These insights are then later helpful in constructing your problem tree and eventual Problem Statement.

In Communication and Media Studies research, breakup and love letters are great tools for understanding specific communication experiences, media content, or tech devices, uncovering the ways in which diverse groups of people enjoy them or are frustrated by them. As with the other Empathize methods covered in this book, breakup and love letters help participants to remember important details that may go unmentioned in traditional methods like interviewing.

What You Need

The tools needed for breakup and love letters are pretty straightforward. When in-person, you can provide participants with pens, pencils, paper, notebooks, and laptops. Or, they can choose to use their own tools. Virtually, you can meet with participants over software like Zoom and have them complete their letters on a computer and add them to a shared Google Drive folder or email them to you.

In any case, you will want to provide some examples. It is likely that your participants have never taken part in this type of study before, so they will need a little guidance to ensure you get the type of letters you are looking for. Sample letters should be an unrelated product or topic. This way participants get an idea of the content and tone for their letter, but no related specifics that may bias their submissions.

Analyzing the Results

Your takeaways after completing this method include the actual letters your participants write, as well as any notes you have taken based on tone and body language while reading their letters. Now you can put your thematic analysis skills to work. You should begin to notice what themes pop out as you read, and reread, each breakup letter and the associated notes. Participants may focus on certain tools or functionalities. Or, they may make mention of experiences or branding issues. You may find that there are a few themes, speaking to more than one "breakup" issue. This is great! More than one core issue means you can create more than one problem tree and attempt to solve more than one UX issue. However, before generating multiple problem trees, do think about what your themes may have in common. UX problem statements can sometimes speak to overarching company issues, while some are just specific tool and functionality issues.

Analyzing love letters can prove a bit more challenging. While breakup letters clearly lead to issues that should be addressed, love letters are praising products and experiences. It seems like, "if it ain't broke, why fix it?" But, instead of thinking of love letters as helping you "fix" something, think of them as helping you be proactive and keeping your app or website up to date. Data need not always be negative; learning what pieces keep users around allows you to extend those positive attributes into other tools and processes.

As one example, let's imagine you are studying Amazon and have asked your participants to write love letters to the digital storefront. But, when you get the data back, you realize analysis is difficult: Everything seems great, so what now? First, take this method in context. Of course, everything seems great, you have asked your participants to write a heartfelt letter praising the product! Second, you still want to rely on your thematic analysis skills. What trends pop out of the letters and associated notes? Instead of thinking just about what specific things users love about Amazon, think about *why* it seems they love these things. For example, here are some snippets from example Amazon love letters:

> Your pricing is always the best and I have rarely ever had to make returns, you have been there for me and have been easy to get a hold of whenever I need you.
>
> You are always there when I need you, no matter what time of the day or night it is!! I know I can rely on you to show me items that I love with prices I can afford!
>
> You are the love of my life, Amazon, because you know the way to my cheapskate heart—amazing prices on products I need and want.

What are some themes you notice right away from reading these three snippets?

In each of these three examples, participants comment on Amazon's pricing. The *why* is fairly clear—people don't want to spend more money than they have to. But there is also an implied trust that Amazon is the best price and perhaps even a hint that participants have compared prices previously and are fairly certain that Amazon's pricing is often the cheapest.

So, let's plug "pricing" into our problem tree trunk in the next stage, Define, and think about the causes and effects of Amazon's pricing model. Following the Design Thinking process, we may eventually find that users would appreciate comparative pricing pulled from other, competing websites, so Amazon customers could more easily compare prices, products, and shipping fees. Users were often already doing this, so adding it would mean less work for customers as well as a brand lift for Amazon—this choice would also show that, as a company, Amazon is confident they can provide lower prices most of the time. As you can see, the pricing was not a "problem," but it could lead to potential product changes that may keep more users and even usher in new users.

Case Study

In February of 2018, Danielle Dennie and Susie Breier, from Concordia University, used breakup and love letters to study how undergraduate students experienced their Library Research Skills Tutorial. The site was created to help students begin their university-level research journey by showing them how to find relevant information, evaluate it, and use it well. The researchers gave students 20 minutes to write their letters. Through their analysis, they learned if students enjoyed the website (most did not). But, perhaps more importantly, Dennie and Breier also learned how students feel about the research process and if their site was fulfilling student needs.[2]

You can see their PowerPoint presentation here! https://spectrum.library. concordia.ca/985363/1/breier-dennie-forum-2019-final-20190426.pdf

Steps

1. Decide what you will be studying.

    a. Is it some experience?
    b. Is it a particular product?

2. Decide what types of letters will be written. Remember to keep an open mind at this point and not let your personal experiences or biases drive this decision.

    a. All love letters?
    b. All breakup letters?
    c. Half and half?
    d. Participants get to choose?

3. Decide if study will take place in-person or virtually.

    a. If in-person, begin organizing a schedule and inviting participants, aiming for five to ten participants per focus group session.
    b. If virtual, decide on the software you want to use. We recommend a virtual focus group via Zoom and a shared Google Drive folder where participants can write their letters through Google Docs.

4. Meet with participants and provide necessary instructions and example(s).

    a. If in-person, provide pens, pencils, paper, notebooks, and/or laptops.
    b. If virtual, walk participants through the Google Folder, making sure everyone has access and knows how to add their own Google Doc. As an alternative, you could set up a private Doc for each participant before the focus group starts, so they can just go into their own Doc and begin typing.

5. Give participants some time to ask questions and review the example(s).
6. Now it's writing time. It is best to give participants about ten minutes to write their letters. This should be ample time to get good content without allowing for any overthinking.
7. Once the writing is complete, ask each participant, one at a time, to read their letters to the group. Take note of their emotions, tone, and body language. If virtual, you may have to ask that a participant turn on their camera. In most instances, you can't force this action. So, if the participant wants to remain a black square, you can still take note of their auditory cues..
8. Allow participants to chat about each other's letters. Remember that the main reason to conduct traditional focus groups is for the interactions that take place. Be sure to take notes of things that participants may include, agree on, or even argue about.
9. Ask any questions that you feel are relevant and would be helpful to better understand participants' experiences, love, and frustrations.
10. Be sure you collect all letters whether they are hard copies or digital copies.
11. Thank the participants for their time, and end the focus group sessions.
12. Begin your thematic analysis, reading over the letters and related notes multiple times. Try to formulate two or three themes that are common across many of the letters.

Discussion Questions

1. Have you ever written a love letter or breakup letter to a person? How do you think this method is similar to, and different from, writing a letter to a partner or ex?
2. Try this method with a group. Pick a product and half of you write a love letter while the other half write a breakup letter. What was difficult? What was easy? How would you use what you learned by being a participant to help guide your participants if you were to employ this method?
3. What could you do if, when asked to write a love letter, a participant states that they cannot think of one thing positive to write. Or, what if, when asked to write a breakup letter, a participant says that they cannot think of one negative thing to write?

Notes

1. "Smart Design: The Breakup Letter," *vimeo*, 2010, https://vimeo.com/11854531.
2. Danielle Dennie and Susie Breier, "The Wind Beneath My Wings: Falling in and Out of Love with an Online Library Research Skills Tutorial," *Concordia Library*, 2019, https://spectrum.library.concordia.ca/985363/1/breier-dennie-forum-2019-final-20190426.pdf.

12 Contextual Inquiry

Contextual inquiry, at the simplest level, combines interviews with observations in context—that is, in the user's workplace, home, or other space where they would use the particular product (website, app, technology, device, etc.) naturally. Watching and listening to a user in their own environment provides insight into how people actually use tools, devices, and platforms and how these are integrated into daily/work routines. Contextual inquiry is very similar to ethnography in that it is more natural and less formal than lab work and does not follow a set task list or rely on scripts. In contextual inquiry, you do not ask the participant to complete tasks, like you would in usability testing. During contextual inquiry, you as the researcher ask questions while the participant navigates a website or an app as they would organically. You come up with new questions based on what you are observing, to best understand how and why the user is doing what they are doing.

Things you can learn during contextual inquiry include: How is a workspace set up? What equipment is used and how it is used? Does the user prefer to use a mouse or keyboard shortcuts to navigate? What issues does the user encounter? How long do different tasks take? Does someone (or the system) help the user navigate when they hit a roadblock? Why does the user do what they do with a digital product?

Quick Tips	
Tools	Recorder (camera ideally), note-taking tools (paper and pen); Zoom or other video conferencing program (for virtual sessions)
Use When	You want to understand how a person moves through a particular routine or flow in their natural environment, typically a work environment. Your research project is about context-specific behaviors and routines or cultures. In Media and Communication Studies, contextual inquiry is particularly well suited for short ethnographic studies of media organizations, such as news rooms or TV stations, or communication contexts, such as classrooms.
Design Thinking Stage	Empathize

DOI: 10.4324/9781003181750-15

The Method

As the name suggests, contextual inquiry consists of observations in a natural environment (in context) while also asking questions of participants about what they are doing and why (inquiry). In this way, contextual inquiry combines the two most commonly used qualitative methods: observations and interviews. Importantly, in the UX field, contextual inquiry is seen as a co-creation method, so it is less about the interviewer asking questions that the participant answers and the researcher then interprets but more about the interviewer and interviewee interpreting the task together. Contextual inquiry is particularly useful for understanding how people conduct activities in the way they normally would, such as their typical behaviors and practices in a work environment, and for thoroughly understanding the point of view of the user.

Hugh Beyer and Karen Holtzblatt developed the contextual inquiry method as a way to better understand workflows and processes in organizations. They came up with this method to overcome the limitations of other qualitative methods, such as interviews and qualitative surveys, which collect after-the-fact, attitudinal data (what people think), but not in-the-moment or behavioral data (what people do in the moment).[1]

A significant strength of contextual inquiry is that it does not only rely on a participant's recall of their actions (which is subject to errors and bias), because it also includes observations of in-the-moment activities. Contextual inquiry also helps the researcher understand the reasoning and motivations for certain actions during an activity, by including interviews with participants as they work. Not only is contextual inquiry great for understanding activities in the moment, it also allows researchers a glimpse into naturally occurring activities, unlike the artificial setting of a lab. Because the activities in contextual inquiry unfold naturally, the researcher also sees all the unconscious, habitual parts of a workflow that a user might not be aware of enough to self-report. For example, a researcher watching someone working with their Word Processor can see if the user saves their progress using mouse clicks in the edit bar or keyboard shortcuts, as saving work regularly is a largely unconscious activity for those working with Word Processors on a daily basis. Observations like this can guide the development of intuitive software interfaces.

Contextual inquiry, as is the case with other qualitative methods, is most powerful when a researcher observes and interviews multiple people working in the same context on similar tasks and then looks for trends or patterns across these interactions. Contextual inquiry is great for understanding the in-depth thinking processes of a user, as well as the unconscious habits and structures they adhere to in a workflow. It is less relevant in scenarios where you're more interested in how a system works and whether it is generally usable.

This method is perfect for centering humans in the design process, aligning well with the Design Thinking Mindset, because it assumes that participants are the subject matter experts in their own lives and that the researcher can learn from observing and asking them questions. Another way of thinking about contextual inquiry is as a craftsperson/master-apprenticeship model, with the participant being the craftsperson sharing the knowledge and experience with their work, while the researcher is the apprentice observing and asking questions to learn more about it.

Contextual inquiry consists of four grounding principles:

1. Context—for contextual inquiry to be true to its name, users have to be observed and questioned working in their natural environment, where they *typically* work, and not in a lab or other artificial setting.
2. Partnership—the session is not guided by the researcher in the way an interview or focus group would be. The researcher should ask initial questions, but the participant should also feel free to guide the conversation in ways they deem necessary for the researcher to understand their work. In this way, the research insights developed are the product of a partnership, a co-creation exercise, rather than an interviewer guiding and extracting very specific, predetermined information from the participant.
3. Interpretation—typically, a qualitative researcher will interpret what they see or hear during the course of collecting data and make recommendations accordingly. In line with the partnership principle, however, contextual inquiry is all about *shared* interpretation (between the researcher and the participant) of the work processes being observed.
4. Focus—it's important that the researcher decides ahead of time what the purpose of the contextual inquiry will be and that they are guided by this focus throughout the observation and conversation.[2]

Contextual inquiry can be conducted either in-person with the researcher observing in the user's physical space or remotely with cameras set up to screen the most important action to the researcher, while the researcher calls in in real-time to talk to the participant. However, during remote contextual inquiry, you might miss some important nonverbal or contextual cues (e.g., people walking past the office who distract your participant from the task at hand)—so it's preferable to do contextual inquiry in-person whenever possible.

One possible problem in contextual inquiry is the introduction of bias. You as the researcher cannot be fully objective. (Objectivity as a human being is not possible!) We all have preconceived notions about certain topics based on past experiences and interactions, but be careful about biases you are bringing with you when interpreting what you see. It's also difficult to make sure that the participant is actually acting naturally. Think about how you would go about your daily activities if you knew somebody was watching your every move! It's impossible to eliminate bias from your research, but being aware of it can help you make more nuanced interpretations of the collected data.

Use When

Contextual inquiry is typically used during the Empathize stage in UX, to understand user workflows, processes, and interactions with various digital products in a real-life setting. Insights gleaned from contextual inquiry can help designers design more intuitive products that fit into how people already work and so make people's jobs easier.

Contextual inquiry can be categorized as a type of field research or ethnography when thinking about it from a more academic research sense, as it combines observing and interviewing in a person's natural environment. It is useful for understanding Media and Communication Studies settings, such as newsrooms, TV stations, and classrooms.

What You Need

You don't need any fancy equipment to conduct a contextual inquiry. As with standard observations, you should have notepaper and a pen/pencil handy to jot down observational notes, as well as a couple of recorders—ideally cameras, so you capture both video and audio—to record the participant's answers to your questions. If you are working virtually, you can use video-conferencing tools like Zoom. Participants can share their screens, and you can record the entire session. Another popular way for participants to "share" their digital screens is the "laptop hug." Here, participants are on Zoom with you on their laptop but want to "share" their mobile phone screen. They then hold their laptop backward, in their lap, so that the screen and camera face outward. Then the participants can hold their phone fairly normally, but you can see what they are doing because the laptop's camera is providing a first-person perspective.

Analyzing the Results

As discussed earlier, contextual inquiry can provide insights into habitual behaviors of users that can guide the design of particular features in product development. Notes and recordings should be thematically analyzed, and you should pay particular attention to pain points or frustrations in the process. This way, you can be ready to move into the Define phase and construct a Problem Statement to guide the design of new or improved digital tools related to the participants' workflows or daily routines!

Case Study

Bryan Dosono, Jordan Hayes, and Yang Wang, all Syracuse University researchers, employed contextual inquiry to explore the experiences of people with visual impairments during the authentication processes of logging into computers, mobile phones, and websites. The researchers visited their participants, all people with varying visual impairments, where they usually use their devices—their homes, workplaces, public libraries, etc.[3]

During their contextual inquiries, the researchers watched as their participants attempted to find login fields and enter usernames and passwords on programs like email, banking websites, social media, and mobile phone operating systems. Most participants used assistive technology, such as the screen readers, JAWS, and ZoomText (screen readers provide speech output for all elements of a website page). After analyzing their contextual inquiry findings, Dosono et al. found that their participants had issues finding where to log in and how to authenticate once a username and password were entered. Screen readers, like JAWS and ZoomText, read from top to bottom. Most programs contain a lot of text and graphic content at the top of the screen, which means it takes users of JAWS and ZoomText a long time to find where to actually enter their credentials. Even when they did find where to log in, the participants were often unsure if authentication was successful.[4]

A key takeaway is the limitations that the researchers found with assistive techs like JAWS and ZoomText. The programs mask passwords, lack feedback when entering case-sensitive passwords, provide little output regarding error messages, and make password recovery difficult for those with vision impairments. On the basis of these findings, Desono et al. provided four suggestions. First, they implored digital designers to improve the accessibility of the login areas, moving them to more obvious positions. Second, they suggested more obvious confirmation messages when successful authentication had occurred. Third, the researchers argued that JAWS, ZoomText, and other assistive technologies should include keyboard shortcuts that help users jump directly to the login areas. Fourth, Desono et al. argue that digital design standards should include consistent terminology for labeling login and authentication areas.[5]

As you can see from this example, just through meeting with 12 digital users in their natural places of digital usage, the researchers uncovered rich data using contextual inquiry. They were able to really experience what users with visual impairments experience, noting many snags and frustrations that likely would have been left out of traditional interviews or qualitative surveys. The researchers were able to produce data-driven suggestions that would not only help visually impaired users but likely make the interfaces more user-friendly for other users as well.

Steps

1. Decide what activities and who you want to observe (define your focus).
2. Decide if you will conduct the session in-person or virtually.

   a. If in-person, schedule a time (or times) for the contextual inquiry with the participants in their natural environment. A contextual inquiry can be as short as a couple of hours or as long as a couple of weeks of observing different activities. Make sure you find a spot in the physical space where you can be unobtrusive but also where you can observe the action.

   b. If virtual, schedule a time (or times) to meet over a service like Zoom. The participants should still be in their natural environments, but you will be "with them" on their laptop, desktop computer, phone, or tablet.

3. Set up any recording equipment (if you can, record from multiple angles). If you are conducting the study virtually, you can use Zoom's recording feature or ask the participants to do a "laptop hug," as explained earlier in this chapter.
4. Start off the session by breaking the ice—introduce yourself and build rapport with the participant.
5. Next, clarify what the purpose of the session is with the participant, telling them what exactly you will be observing and detailing how the session will proceed. Make sure to explain that you'll be largely observing and asking questions as they arise. Also clarify with the participant when it's OK to interrupt and when it's not.
6. Start the contextual inquiry. The majority of your time should be spent watching the person go about their activities. Make sure you maintain focus on understanding what you're seeing to address the Problem Statement guiding your research.
7. Stop the observation, and ask questions when you don't understand what the participant is doing or why. Ask them to walk you through their actions and motivations and to provide lots of details!
8. Continue the contextual inquiry with alternating periods of observation and periods of discussion. Make sure to take copious notes!

   a. When you make interpretations about a particular part of the work process, validate these interpretations with your participant (remember the partnership principle)—but don't overwhelm them with interpretations in the moment; you can confirm your interpretations toward the end of the session. Providing interpretations continuously and probing in particular directions during the session can cause the participant to change their behavior to please you.
   b. When observing and asking questions, get information on whether what you're observing is the standard way of doing things or if it's an anomaly (and if it's an anomaly, ask why it is occurring).

9. At the end, summarize the session and share your interpretations with the participant. Ask them to weigh in on these—they might need to clarify or correct some of your interpretations.

Discussion Questions

1. Think about some specific scenarios of when you would conduct a contextual inquiry rather than an interview. Why would a contextual inquiry make more sense here? In what scenarios would a contextual inquiry *not* make sense?
2. How could you make sure that you are unobtrusive in your observations in a physical space? Provide some practical tips.
3. What may be better captured in-person during a contextual inquiry than virtually? Can you think of any examples of data that would actually be captured better during a virtual session and that may be missed during an in-person session?

Notes

1. Karen Holtzblatt and Hugh Beyer, *Contextual Design: Defining Customer-Centered Systems* (Amsterdam: Elsevier, 1997).
2. Kim Salazar, "Contextual Inquiry: Inspire Design by Observing and Interviewing Users in Their Context," *Nielsen Norman Group*, December 6, 2020, www.nngroup.com/articles/contextual-inquiry/.
3. Bryan Dosono, Jordan Hayes, and Yang Wang, " 'I'm Stuck!': A Contextual Inquiry of People with Visual Impairments in Authentication," in *Eleventh Symposium On Usable Privacy and Security ({SOUPS} 2015)*. 2015, https://www.usenix.org/conference/soups2015/proceedings/presentation/dosono
4. Ibid.
5. Ibid.

13 Personas and Scenarios

Personas are all about designing for a specific user in mind. Personas are fictional representations or profiles of different target users that a researcher creates based on information gathered during the Empathize stage. So, personas are not a data *collection* method but more a data *analysis* method—they are a way of organizing your Empathize findings to be used throughout the iterative design process. Personas, also called model characters, should represent an "ideal" user and should help you categorize possible different user types in simplistic ways, to aid design.[1]

Scenarios work in tandem with personas. They are fictional narratives of how, when, and in what context a typical user would visit your digital space. Scenarios can be either goal/task-based or more elaborated, delving deep into the ideal user's motivations and circumstances. Scenarios can be used with personas to help UX designers better understand the end-users that they're designing for—it's easier to design for a name, face, personality, and story than for a generic "target user"! Think of personas and scenarios in the language of filmmaking: Where the persona is the main character in the story, the scenario is a particular (typical) scene in the movie (one in which the character interacts with a digital product).[2]

Quick Tips	
Tools	Pen, paper, software for document creation
Use When	You want to categorize and organize user data collected during the Empathize stage into different user types to shape your design. You are categorizing and analyzing qualitative data, particularly when you're interested in commonly-found differences between groups of people in terms of their needs and experiences.
Design Thinking Stage	Define

DOI: 10.4324/9781003181750-16

The Method

Personas are fictional representations or "hypothetical archetypes" of potential users that can help guide design decisions. The concept of personas was developed by Alan Cooper, a software designer in the 1980s. After conducting some interviews with his target users, Cooper decided to play-act as imagined characters loosely based on the user information he had gathered when brainstorming the development of new software. He found this role-playing an effective way of putting himself in the user's mindset and representing different points of view in the design process.[3] And, personas were born! For more on Cooper's fascinating journey with personas, check out his essay (it's in the references of this chapter).[4]

There are different ways of developing personas, depending on the needs of a project. Traditional personas are in-depth, fleshed-out profiles that categorize different user types and that answer the question "who are we designing for?" (as well as connected questions, such as, how and in what context will these users be interacting with our product?). Personas can be created in two ways: from secondary research (existing documents, previous research findings) and from Empathize research for a specific project. Basically, before you develop personas, you need to know something about your targeted users—be it from published studies and asking subject matter experts or from speaking to possible users themselves (in the form of interviews, focus groups, screenshot diaries, etc.).

A persona is an ideal, reliable, and realistic but fictional representation of a specific target audience segment. Personas are archetypes of users, highlighting their key characteristics and should represent different types of possible users who might be interested in your product. Personas are written, as a one-to-two-page document, by UX researchers and designers, and look like a biography or profile page. Personas typically include:

- Picture (use stock photos or illustrations)
- Name (use a fictional name)
- Demographic information, such as age and gender
- Background information, such as location and occupation
- Goals and desires (related to the product)
- Frustrations and pain points (related to the product)
- A typical day
- Highlights of the persona's personality, behaviors, skills, and habits

Personas should include rich fictional details, such as quotes and specific details that are relevant to the context of the digital product. For example, personas developed for a dating app should focus on relationship characteristics, such as what a person wants from a partner, whereas personas developed for a health app might focus on things like nutrition and exercise habits.

An important thing to remember about personas is that they should not be "made up" but should be based on some actual data about your possible users, whether it's secondary research or data you collect yourself. In addition, personas should be connected to a particular context, not just be random characterizations.

Figure 13.1 is an example of a persona related to a (fictional) new coffee delivery app, Cuppa Joe Quick Coffee, that delivers coffee quickly to specific pick-up points.

Personas like this give a lot of insight into our target users. Because complex personas provide a lot of information and can take a long time to develop, it is sometimes easier to create lean personas, which are essentially more concise versions. These personas present only the most pertinent information and are developed from a few (five to ten) quick user interviews conducted just for this purpose, rather than lots of background and previous user research.

Some questions to ask in lean persona interviews include:

* How do you spend a typical day?
* How do you use [a particular tool or product] during your day?
* What parts or features of it don't work for you and why?

If you don't have time to do even a few interviews (as is often the case!), you can also do some online research, validate sample personas with people in your network (students, co-workers, family, etc.) who match the profile, and check out competitors to understand better what types of people they are targeting. Sometimes personas are created based on assumptions held by the

Keisha (she/her/hers)

Digital native
Good at multitasking
Likes to eat healthy and exercise
Has trouble sleeping

Age: 24
Occupation: Marketing Coordinator
Family: Single
Location: Philadelphia, PA USA

A Typical Day

Wake up around 8, start work at 9:30 (depending on what time she wakes up and leaves, she will get either the bus or walk to work - though often runs late), typically has lunch at her desk, finishes work at 5, gets the bus home or to happy hour with her friends. Spends the rest of the evening watching TV with her cats.

Goals• Interests

· Wants to move through her day efficiently so that she has time to relax in the evenings
· Very comfortable with technology, particularly her mobile
· Is saving for a car so sticks to a budget
· Drinks coffee to keep her awake and not for the flavor

Pain Points •Concerns

· Does not like having to make phone calls, prefers texting and communicating online
· Is always running late

Motivations

Saving money
very high

Being efficient
very high

Personality

introvert				extrovert
●	○	○	○	○
analytical				creative
○	○	●	○	○
loyal				fickle
○	○	○	○	●
passive				active
○	○	●	○	○

Figure 13.1 A sample persona.

UX team about ideal users: Be careful when not basing your personas on real user data, as you might fall into the trap of creating stereotypes instead of archetypes.

Creating personas is a bit of a balancing act—you don't want to add too many details that aren't relevant, but you need enough details that allow designers and product teams to empathize with the target user and understand their point of view. Typically, in UX, the team creates multiple personas per product, as there are usually slightly different audience segments that you want to appeal to (there is no one monolithic target user). For instance, if you are creating an app for time management for parents, you might have one persona representing new parents of their first child and another one for parents of multiple, older children.

How many personas should you create for a product? While there is no "right" number of personas, three to five per product is a good number to aim for. This is enough to allow designers to consider possible variations in behavior and attitude in the target audience, while also not being too overwhelming. A way of organizing personas is to designate one as a "primary" persona that the design should be geared toward, but then test the product (by validating with users with similar characteristics) with the other personas created for that product.

After you have created some personas, you should create realistic stories or scenarios for them regarding how, why, where, and when they are likely to use your product. Scenarios describe a typical imagined interaction with a product. Here is an example of a scenario:

> Keisha (our persona from earlier in this chapter) runs late to work a lot. She has trouble sleeping and a hard time waking up, which means she often rushes through her morning routine. To make matters worse, her roommate Riley frequently uses up the coffee beans that they share and rarely replaces the bag. Keisha needs coffee so she has enough energy to get through her busy day. But she's typically running late already and has not much time to spare, so she is looking for the most efficient way of getting her caffeine fix. She is looking for ways to order coffee regularly online, with minimal effort and time, to pick up on her walk to work. She sees Cuppa Joe Quick Coffee advertised on her Instagram and decides to explore the app, to see how it can help her mornings run more smoothly.

As with personas themselves, scenarios need to strike a balance between providing enough detail to elicit empathy but also being succinct and relevant to the context at hand. It's a useful exercise to build your scenarios around particular goals that the user wants to achieve using your digital product (e.g., look for ways to order coffee regularly online for pick-up, as in the scenario above). Remember to include the motivations for why this persona wants to achieve this goal, to inspire the design!

Use When

Personas are typically created in the Define stage, after you have collected some user data in the Empathize stage. Personas are used by design and product teams to focus on who they are designing for, because they highlight the motivations and expectations of ideal users. They help designers build empathy for their users and understand their perspectives better. This helps to prioritize certain features that talk to specific user needs. Personas are great for getting designers to design for others, not for themselves. (It's easy to base decisions on what *you* would do in a particular situation—personas force designers out of their skin.)

Personas are essentially a way of analyzing qualitative data, so they are similar to thematic analysis in academic research in a sense. To create personas, a researcher will go through data collected from users or secondary data about users and look for themes and patterns in order to create personas of different user types. Personas are useful for working with group differences in academia research. When going through their collected data, researchers can create a tangible character and story for themes found around group differences. In this way, personas can be a useful and creative reporting technique in academic research, humanizing findings based on themes found in the data.

Personas are also useful, both in UX and in academic research, for engaging with users in further research. You can show users personas you've developed to validate whether the insights and themes you've come up with resonate with users and whether these personas seem actually representative of themselves or people they know. They are also useful as exercises in getting participants to put themselves in others' shoes and imagining different scenarios, as a method of understanding attitudes and opinions around various topics.

What You Need

As with most UX methods, you do not need fancy equipment to develop personas and scenarios. You need the documents—transcripts, notes, summaries—from any previous data (whether secondary data that others collected or primary data from research you did with users) and a way of looking through these documents systematically—this typically involves a Word Processor on a computer where you can have multiple documents open and can color-code and comment on the content to distinguish between your various patterns and themes. You can create a simple persona in a Word or Google doc, using text and image formatting functionalities of the programs.

Case Study

The UX team for Spotify, the music streaming app, developed personas in order to better understand their current and potential users. An app like Spotify is technically "for everyone"—all types of people listen to music—so using personas can really help designers to focus on the needs of different listeners (to differentiate mentally between users), and not have to design for some ambiguous mass audience. Because personas are an internal method, Spotify does not share the actual personas themselves (with all the pictures and characteristics) outside of the company. So even though we can't provide Spotify's full persona details here, we still see value in sharing the persona development process and how it was used by the company.[5]

The team at Spotify started the persona development process by analyzing their current user base. The UX researchers conducted diary studies and contextual inquiries with listeners of different ages, lifestyles, etc., to better understand why people listened to music and their needs. Early on, the team realized that the reasons that people listen to music are actually pretty similar across groups (e.g., for entertainment, to kill boredom) but that there were key differences around device use, the contexts (how and where people listen to music), and whether people were prepared to pay for music or not.

In the next phase of the project, the team decided to hone in on better understanding how people listen to music together. They went through the user data again, specifically focusing on contexts where people listened together. They then created five personas, each representing a person who listened to music with others in different ways and spaces, such as at parties and on commutes. The team picked genders, names, and characteristics of the personas randomly, based on the range of actual users from user research, and created cartoon pictures for each persona. Next, the research team shared the personas across the company so that different teams (design, marketing, sales, etc.) could refer to them to guide their work in specific areas. Not only did the team share the personas as static images but they also created an interactive website, ran workshops, and even created a card game so that everyone at Spotify could really understand and relate to the personas in their work.

Steps

1. Collect user data, whether through your own or secondary research. This step includes conducting interviews and observations and parsing through existing documents. It's also a good idea to talk to other people in the organization to put together everything that they know about our users. If you're working on improving an existing product, you can also rely on existing analytics (such as web analytics—information about existing customer behavior on your website, such as number of clicks or time spent on the website) to guide your persona development. This step is really important, so you don't simply create stereotypes of users based on common knowledge. You should collect some real data as a basis for your personas.
2. Once you have gathered user data, you're ready to analyze. Analyzing requires going through the data (interview transcripts, observation notes, analytics, etc.) and looking for patterns, particularly looking for patterns

that highlight different behaviors across all the users. Repeated behavioral patterns should form the basis of each of your personas.

3. The next step is actually creating the personas themselves. It's helpful to start with a persona template, that includes all the details you want to flesh out per persona, such as name, age, occupation, attitudes, skills, etc. (Remember to make your persona context-specific, that is, connected to the digital product you are interested in designing or improving.) Quick tip: Don't base your personas on people you know (use fake names and photos), and also don't simply write out the characteristics of one specific user to serve as your persona. *Personas should be based on a combination and distillation of information from multiple real users.*

4. After you have created personas, you need to create scenarios for them. Personas become much more powerful when they have a story behind them.

5. Once you've created your personas, you should "socialize" them. This means you share your personas and scenarios with the rest of the team, so they can start talking about them and designing and developing the product for these "ideal users" in mind. Personas are particularly powerful in that they provide a tangible focus for designers and engineers as they go about their work—everyone works with an actual (if fictional) user in mind, not just a nameless faceless audience segment.

Discussion Questions

1. Think of your favorite website or app and how you navigate through it. Write a persona based on your behavior on the app, as well as a scenario of typical use, focused on improving this app in some way that you would like to see. Interview some of your friends who also use this app. What other personas and scenarios could you create based on their answers?

2. How might you use personas in creative ways as activities during interviews or focus group participants? Think particularly about diversity, equity, and inclusion and the value of getting people to empathize with others in your research.

Notes

1. Rikke Friis Dam and Teo Yu Siang, "Personas: A Simple Introduction," *Interaction Design Foundation*, January 2021, www.interaction-design.org/literature/article/personas-why-and-how-you-should-use-them.
2. "Scenarios," *Usability.gov*, accessed July 18, 2021, www.usability.gov/how-to-and-tools/methods/scenarios.html.
3. Rikke Friss and Teo Yu Siang, "Personas."
4. Alan Cooper, "The Long Road to Inventing Design Personas," *OneZero*, February 4, 2020, https://onezero.medium.com/in-1983-i-created-secret-weapons-for-interactive-design-d154eb8cfd58.
5. Mady Torres de Souza, Olga Hording, and Sohit Karol, "The Story of Spotify Personas," *Spotify Design*, March 2019, https://spotify.design/article/the-story-of-spotify-personas.

14 Problem Trees

Remember that, as a UX researcher, you are worried not only about desirability but also about feasibility and viability. Your studies and recommendations cannot exist in a vacuum; they should always be linked to the culture and current reality of your app or website, including the company's mission, so it's a good idea to consider these factors when planning a study. A problem tree takes preliminary data from the Empathize stage, as well as previous research, and assists you with recognizing the core issue, outlining the causes and effects, and constructing a clear and concise Problem Statement.

Quick Tips	
Tools	PowerPoint, Word, Illustrator, Gimp, Google Jamboard, Mural
Use When	You are attempting to craft your Problem Statement and need to define the roots and causes of your problem. You are crafting your Research Question and need some help making it clear and concise.
Design Thinking Stage	Define

The Method

Problem trees are visualizations of a core problem with your product. The "trunk" of the tree represents the core issue, the "roots" are the causes of that issue, and the "branches" are the potential effects. Essentially, you are mapping the core problem that you found via Empathize methods and the causes and effects of that problem. Some causes and effects may actually be found through conducting analysis of participants' responses during user research. You will also apply your general/common knowledge to add other possible causes and effects. Creating a problem tree helps you to break down the problem into smaller, more manageable pieces.[1]

DOI: 10.4324/9781003181750-17

It is extremely likely that you will actually formulate more than one Problem Statement. While at first this may seem like bad news because it means that you have to "pick" one or that you have just given yourself more work, in reality, it is a big positive. As a UX researcher, you want to constantly have multiple projects ready to go that provide the rationale for why your job is important and necessary. Also, of course, it's your job as the UX researcher to be proactive, always suggesting small changes to the app or website *before* users become overly frustrated. Depending on the size of your team, you may be working on several projects at once, or you may work on one Problem Statement at a time.

After analyzing your Empathize results, decide what the problem or issue is that needs to be addressed. If you find more than one large issue, you can construct more than one problem ree! The *core* problem can be broad at this point. The process of building the tree will help to narrow your actual *research* problem (or a more specific Problem Statement). Once you have your core trunk, you will map the roots (causes) and branches (effects). Again, these will come from both your participants and your knowledge of the product. You don't necessarily have to map the problem, causes, and effects as a literal tree. But, just be sure the way you map it shows a clear cause and effect visualization. It is also common to have more than one layer of causes and effects. One cause may lead to another, secondary cause. And, one effect may lead to another, secondary, effect.

Here is a sample problem tree for a study abroad website. Let's imagine that, through some Empathize research, you find that participants are having issues moving through the mobile version of the site—notice this is the "trunk" or core issue.

What do you notice after reading through this problem tree? First, take note that the core issue is directly driven by data collected during the Empathize stage. The problem is still vague and messy—it would be difficult to conduct an effective Ideate method at this point. The roots tap into different pieces of the website, including coding, design, and ethical considerations. The possible effects not only list that users may be turned off and not return but also speak to larger branding issues and ethical issues, including ostracization caused by lack of inclusivity.

As you read over some Ideate methods in later chapters, keep this problem tree in mind. What could be some Problem Statements that could come out of this tree? What would make for good Ideate methods to use on this project? Are there any pieces that may warrant looping back and conducting another Empathize method?

Another Way to Define: Task Analysis

A problem tree is just one way to help researchers hone in on the Problem Statement by breaking down and detailing out a larger problem. Another great "breakdown" method that can be used in the Define stage is task analysis, which we'll briefly explain here. In task analysis, the researcher creates a

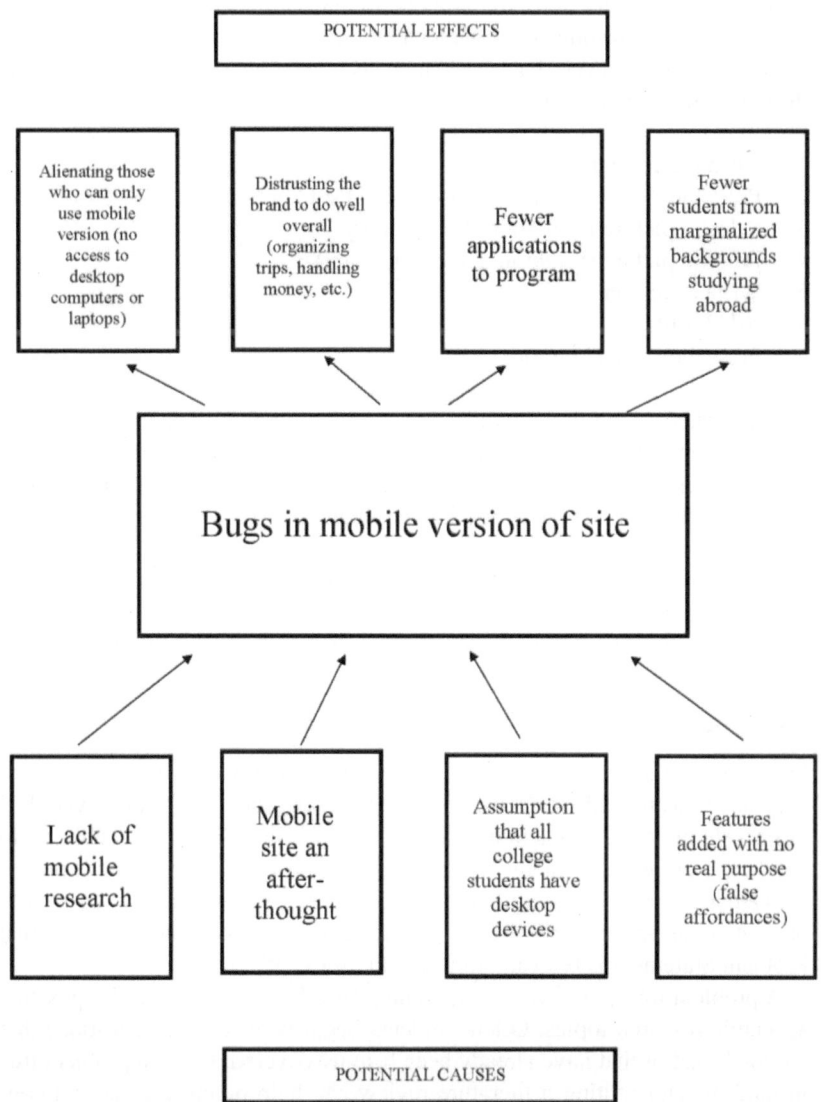

POTENTIAL EFFECTS

Alienating those who can only use mobile version (no access to desktop computers or laptops)

Distrusting the brand to do well overall (organizing trips, handling money, etc.)

Fewer applications to program

Fewer students from marginalized backgrounds studying abroad

Bugs in mobile version of site

Lack of mobile research

Mobile site an after-thought

Assumption that all college students have desktop devices

Features added with no real purpose (false affordances)

POTENTIAL CAUSES

Figure 14.1 A sample problem tree based on a Study Abroad website.

diagram with all the steps and substeps that a user needs to take in order to complete some goal (e.g., all the steps needed to "post a picture" on Instagram). The UX team can then study this diagram to inform their design, figuring out ways to make the task sequence easier and more intuitive for the user. For instance, some of the listed steps might need additional user support built into them or some steps can be eliminated completely by designing a more intuitive product. Once a task analysis is complete, you can validate it—that is, check whether all the steps you laid out are indeed the steps that users take—through usability testing (see Chapter 19).

For example, in a very basic task analysis of posting a picture on Instagram, the steps might go as such:

1. Open the Instagram app.
2. Press the "+" button and choose "Feed Post" option.
3. Select a picture from your phone gallery.
4. Edit the picture (brightness, cropping, etc.).
5. Write a caption.
6. Add hashtags.
7. Press the "share" button.

If you looked at the previous discussion to help define your Problem Statement, you might decide to focus on eliminating some steps to make the flow more efficient. So, for example, you might decide to add a feature that would automatically generate some possible hashtags for a picture, using AI to categorize the picture. To learn more about task analysis, check out Usability Body of Knowledge's page here: www.usabilitybok.org/task-analysis. Or, Interaction Design Foundation's page here: www.interaction-design.org/literature/article/task-analysis-a-ux-designer-s-best-friend. Now, back to problem trees!

Use When

Problem trees are best constructed when some preliminary research is available, and you are ready to decide on your Problem Statement. Nearly every UX project should use a problem tree (or a similar method, like a task analysis) to define and organize problems with the app or website. This is why it is part of the Define stage of the Design Thinking Mindset. It should be used fairly early on, but some preliminary research is necessary to ensure that your eventual Problem Statement is relevant, timely, and user-centered.

A problem tree can also be used to help sort through and organize possible academic research topics. Often, students begin with research questions that are too broad or that have already been heavily covered. Using a problem tree in tandem with writing a literature review can help output a clear and concise Research Question. For example, if you have an idea for your research problem, begin by putting that into your problem tree as the "trunk" or core issue. This can be something broad, perhaps a trend you have noticed in your personal daily life, a friends' experience, or even something you read about or saw on the news. Now, as you conduct a search for relevant literature, you can start to add roots and branches, or causes and effects, respectively, to your tree. By picking one or two of these causes or effects, you will be able to construct a targeted and important Research Question!

What You Need

To begin, you will need preliminary data, most likely from Empathize method results. To construct the actual tree, you can use tools as simple as a paper and pencil or a whiteboard. A whiteboard is especially helpful in the earlier stages of problem tree creation because it allows for constant editing and smoother collaboration.

To formalize the tree, you can use simple digital tools like Microsoft Word and PowerPoint. These software allow for the creation of shapes, arrows, and text boxes, which is all you really need for a problem tree. If you are comfortable with software, and want to get a little fancier, you can use programs like Adobe Illustrator or Gimp. If you need to virtually collaborate to create your problem tree, Google's Jamboard or Mural are both great collaborative whiteboard tools.

Analyzing the Results

Once your visualization is complete, it is time to decide what your Problem Statement will be. Your Problem Statement should be concise and clear. It should be manageable and able to be worked through using your data and processes that you know you have the resources to complete. It should also not only speak to users but also keep in mind the company's goals and mission. You may have already learned about Research Questions. Similar rules follow for Problem Statements, but in UX research, we state the issue instead of asking about it, since we already know there is a problem from initial Empathize research.

Looking at the tree, you can ask yourself, or your team, some questions:

- What are the causes?
- Are they directly related to specific effects?
- What is possible to research?
- How would we research that element?
- Can we attempt to understand one root or one cause and actually fix multiple pieces of the app or website?
- Have others dealt with similar issues? (Read more about competitive analyses in Chapter 6.)
- Are all socio-cultural dimensions of the problem considered?[2]

Good Problem Statements usually come out of a cause (root) or effect (branch). Because the core issue, or the trunk, is often too broad, the problem tree visualization helps us to link roots and effects and hone our research problem. Your final Problem Statement will probably be similar to your trunk but made more precise through at least one cause, one effect, or some combination.

Case Study

It's difficult to find a published problem tree because it is such an internal method. Instead, we will share another sample problem tree. In this problem tree, imagine you are interested in studying Pinterest. You ask participants, during the Empathize stage, to write breakup letters. After analyzing that data, you find that users were experiencing findability issues—they couldn't find specific content on the platform. As you can see in Figure 14.2, listing the causes and potential effects breaks down the issue into more digestible chunks.

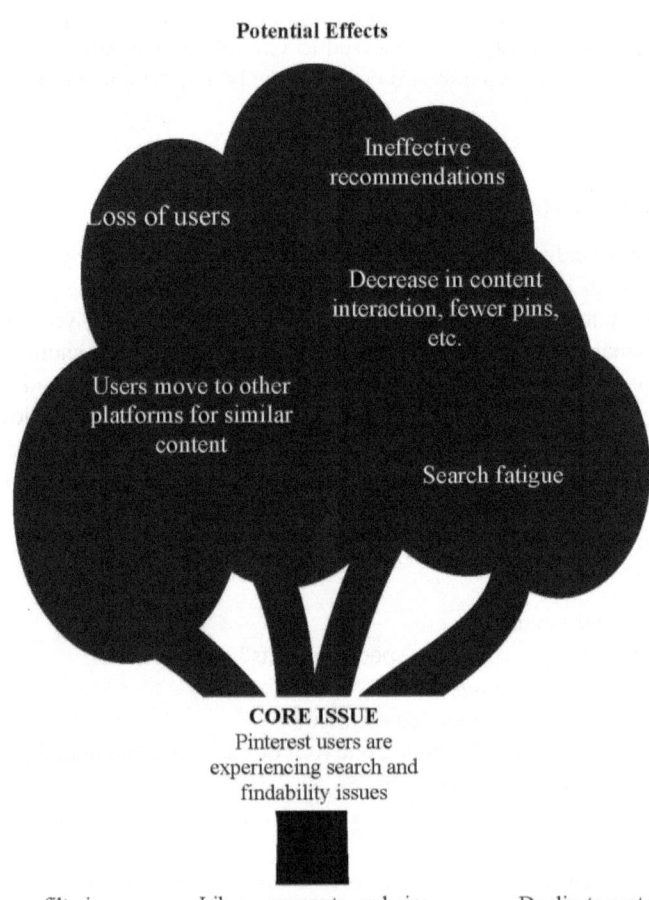

Figure 14.2 A sample problem tree based on Pinterest.

Now your Problem Statement could look something like: "Sorting and filtering options are lacking within Pinterest's search function." Notice this one Problem Statement was derived through one root. There are certainly other Problem Statements that can also be gleaned from this tree. What are some others you can formulate?

Steps

1. Complete your analysis of Empathize method results.
2. Decide what your core issue is. It is OK if it is broad at this point.
3. Make this core issue the "trunk" of your tree.
4. Use your participants' contributions, as well as your own knowledge, to decide what the causes of this issue are.
5. Map these causes as the "roots" of your tree.
6. Again, use your participants' contributions as well as your knowledge to decide what the possible effects of this issue are.
7. Map these effects as the "branches" of your tree.
8. Consider if there are secondary roots and causes. These are causes of causes and effects of effects!
9. Map these secondary roots and causes as well.
10. Edit your tree to see if you can match up some roots and effects, noticing possible linear processes.
11. Analyze your tree to recognize what potential studies may solve more than one issue!
12. Finally, decide on a concise Problem Statement to tackle.

Discussion Questions

1. How is the process of creating a Problem Statement different from constructing a Research Question? How is it similar?
2. Try drawing a problem tree for some process at your college or university. For example, maybe there is limited parking for students. Or, maybe the food on campus just isn't great. What are some potential roots and causes for this problem? Do any solutions jump out that were not obvious before?

Notes

1. e.g., Ingie Hovland, *Successful Communication: A Toolkit for Researchers and Civil Society Organisations* (London: Overseas Development Institute, October 2005), https://citeseerx.ist.psu.edu/viewdoc/download?doi=10.1.1.123.1057&rep=rep1&type=pdf.
2. See these and more discussion questions in: Hovland, *Successful Communication*, 13.

15 Cognitive Mapping

It is no secret that it is difficult as a researcher to truly understand a phenomenon from your participants' perspectives. So many variables make up how users experience apps and websites. Not only do different people focus on different tools and functionalities when using a product but also users' backgrounds, cultures, ages, locations, races, genders, disabilities, and so on all impact their experience. The outputs of cognitive mapping are visual representations of specific users' experiences—what they remember about an app or website, what is important to them, and what they wish the product included.

Quick Tips	
Tools	Paper and pencil; Zoom, Skype, Google Jamboard
Use When	You want a visualization of what users remember about an app or website, their interpretation of the space, or how they would change an interface or process, so you can incorporate these perspectives into your design. Your research question is about wanting to see the "picture in people's heads" about a particular concept, object, or process.
Design Thinking Stage	Ideate

The Method

It is commonly understood now more than ever that people learn in different ways. Guidelines like Universal Design for Learning (UDL) focus on the varying ways people best engage with, represent, and express new information.[1] In particular, guides strongly suggest that people are offered multiple modes of expressing learned content, including traditional writing, creative essays, performances, drawings, and videos. These ideas about learning are not only beneficial in education, they are also extremely helpful in qualitative research.

Traditional academic methods in Communication and Media Studies include surveys, focus groups, or one-on-one interviews. These methods ask people to

DOI: 10.4324/9781003181750-18

either talk about some experience or write about it. But, these types of outlets are not always the best ways for people to express their emotions, feelings, needs, desires, or frustrations. Cognitive mapping is a creative approach that instead asks users to create sketches, drawings, or mappings of certain concepts or products. In essence, as the researcher, you are asking participants to create mental maps.

The original idea for cognitive mapping can be traced back to 1948, when psychologist Edward Tolman was experimenting with brain mapping, mazes, and rats, showing how rats create mental maps of their environments in their heads.[2] In the late 1980s and early 1990s, cognitive mappings started to become a popular method in social and behavioral sciences.[3] For instance, cognitive mappings are popular in geography and social work studies fields. Researchers ask their participants to draw physical spaces like city blocks and the inside of buildings to better understand how people move through physical spaces, what they remember and thus are important pieces to them, and how included or excluded they feel in certain areas.[4] Interestingly, cognitive mapping works well as an app and a website method because digital spaces are similar to physical spaces. Users "move" through digital spaces, utilizing different tools and feeling different emotions based on who they are, their cultural backgrounds, and their relevant goals as they go.

Cognitive mapping is also a great method to really involve your participants in the Ideate process. Instead of asking participants to look at a current interface or process and then comment on what they may change or where they feel excluded, you can provide a blank slate. Beyond asking participants to draw what they remember from a current version of an app or a website, you can also ask them to draw how they wish an interface looked or how a process is organized. Without the current version of your product to act as a guide, participants are more likely to provide authentic feedback that truly captures what they would like to see, not what they think the app is supposed to be doing based on the current version.

In addition, you can also ask that participants draw a different, but related, product that they like better. For example, perhaps you work for Pandora, the radio station app, and you have found that users think the way stations are organized is confusing. You may ask that participants draw a different music streaming app that they believe provides a more intuitive organizational experience. From these results, you are not looking for exact changes to your product to be perfectly visualized, but rather you want to understand what it is that another product provides that yours does not.

It is up to you to decide on how much or little guidance you will give to your participants. This is, of course, guided by your Problem Statement or Research Question and the goals you are trying to reach with your study. You can leave it up to them to draw, sketch, map, or chart however they please. Or, you can specifically ask for a sketch or flowchart.

As with many methods discussed in this book, a large part of a great cognitive mapping study is asking good questions. Cognitive mapping can be conducted virtually or in-person. But, in both cases we recommend a synchronous meeting. As your participant draws, you can ask that they think aloud, explaining what they are including and why. Then, after they have completed each drawing, you can ask them questions to further explain their mappings, taking notes or directly annotating their images.

In-person cognitive mappings can be done simply with sheets of paper and pencils. You can take notes while participants walk you through their drawings as well as annotate their mappings once they have completed each one and handed it to you. Virtually, tools like Google Jamboard or Mural (both digital whiteboards) paired with video communication software like Zoom work well. Speaking with your participants through Zoom, you can view a shared Google Jamboard and watch as the participants "draw." You can then easily save these boards for later analysis, including the notes and annotations you added.

Use When

For this textbook, we are suggesting cognitive maps be used in the Ideate stage. Cognitive maps can also be an Empathize method, when you ask participants to draw current systems or processes. But because mappings can provide creative looks into how participants would change a product or help solve a research problem, we like to think of them as an Ideate method. Remember, as UX researchers, we are *not* the users. Thus, cognitive maps provide unique insight when working to decide what a prototype will include. Cognitive maps are best used in the Ideate stage with prompts like "draw what would you change about this product" or "draw another app/website that includes a similar feature but does it better."

However, cognitive maps can also be used in the Empathize stage. Instead of asking participants to draw potential solutions or changes, as discussed, you can instead use cognitive mapping to tap into how users feel about your product currently. At a macro-level, you can ask that participants just "draw" your product. This straightforward prompt offers insight into how users view your website or app on two levels. The first is through tools and functionalities—noting what your participants include and omit speaks to what they find important and memorable. The second is more symbolic—what users include in their mappings can be analyzed to provide a deeper understanding of their thoughts on the space.

As one example, take some time to view the two sketches of Twitter in the following. Let's imagine you provide participants with the simple prompt: "Draw Twitter." What do you notice about the two contributions in Figure 15.1? How are they similar? How are they different?

In the two sample sketches in the figure, it is clear to see that both include a few of the same elements—which are the main pieces of Twitter—tweets and hashtags. But we can also see that the first is mainly focused on content, not giving much attention at all in their drawing to the actual structural or aesthetic elements of the app. The second, however, includes more structural elements, including profile photos, the settings gear, ads, suggested accounts,

Figure 15.1 Two sample cognitive mappings of Twitter.

and the general organizational aesthetic of the Twitter feed. These findings, paired with interview questions and demographic information, would begin to provide insight into what is important to users.

The example in the figure would more likely be the result of an Empathize method, or a Research Question that is attempting to understand how users feel about, or experience, an app or an website. On the other hand, cognitive mapping is great to use in the Ideate stage. Let's imagine that, through your Empathize stage, you found that participants were not posting to Twitter and had some frustrations with the posting process. You may then prompt your participants with: "Draw an app's posting interface that you enjoy." Findings may vary. For instance, participants may all sketch different apps. But, remember, after collecting cognitive mappings from your participants, you should not expect exact prototype changes or Research Question answers to pop out. Instead, a thorough analysis is needed first.

What You Need

In-person, the cognitive mapping method requires just a pencil and some blank sheets of paper. If you are conducting the study virtually, you will need a shared digital drawing space, as well as some place to synchronously connect. We suggest Google Jamboard paired with Zoom.

Analyzing the Results

As an Ideate method, the goal of cognitive mappings is to better understand what changes should be made to potentially resolve the research problem. However, you should not expect perfected prototype changes to spill out of your participants' sketches. Instead, analysis should consider the interview data you collected while the participants were drawing along with your annotations. Some questions you should ask yourself during analysis include: What did participants include and what did they leave out? What symbolic features did participants include? Are there any trends that seem popular within certain demographic identifications of participants?

If asking participants to map how they wish your product was designed, be sure not to just take all changes at face value. Often, if asked, users will provide long lists of things that could change. However, this doesn't mean that participants would actually use all of those new tools or could find ways to incorporate them into their everyday lives. This is why asking good questions and critically analyzing users' contributions is so important. Remember, as an Empathize method, you are looking to understand user delights and frustrations. As an Ideate method, you are hoping to gain some ideas that begin the prototyping process.

Case Study

In 2018, Michael A. Devito, Ashley Marie Walker, and Jeremy Birnholtz, all scholars from Northwestern University, used cognitive maps as a method to understand how LGBTQ+ people self-present their identities on social media platforms. Participants were provided with paper and colored markers. They were then asked to map how they present their LGBTQ+ selves online and to be creative when doing so. By employing the cognitive mapping method, the researchers helped their participants to visualize their digital identities and relationships without the overarching platforms themselves getting in the way and on their own terms. The researchers then used the sketches during the interview phase, asking participants to explain what they had included.[5]

Devito, Walker, and Birnholtz found that their participants use social media to both express their LGBTQ+ identities and avoid ostracization and harassment. Because different social media platforms provide different audiences and expected post types, these participants were able to perform the selves they wanted to and needed to based on who was watching. In particular, different platforms, or even multiple accounts on the same platform, allowed for audience segregation to ensure proper identity management.[6]

While this example is a more traditionally academic study, utilizing cognitive mapping clearly helped the researchers gain insight that was unlikely to come out during a traditional interview or qualitative survey. In addition, because participants could sketch whatever they wanted, the researchers were provided with a myriad of images including sliding scales, social media maps, lists, and cartoon representations, which made for a richer, more nuanced analysis.

Steps

1. Decide what you will be studying.

    a. Are you using cognitive mappings as an Ideate method? What prompts will you include?
    b. Are you using cognitive mappings as an Empathize method? What prompts will you include?

2. Decide if the study will take place in-person or virtually.

    a. If taking place in-person, prepare paper and writing implements (pens, pencils, markers, etc.) for sketches.
    b. If taking place virtually, set up Zoom meetings and Google Jamboards.

3. Decide if you would like participants to draw a specific type of map—sketch, flowchart, list, etc.—or if you will allow them to freely map.
4. Once you are sitting down with each participant, explain what creating cognitive maps entails. Be sure you let participants know that drawing ability doesn't matter. Often, participants are worried they are going to be "graded" on the basis of their artistic skills.
5. Before providing the first prompt, be sure the participants are comfortable with the tools provided and understand their role.
6. Begin by providing the first prompt. Be sure you only provide the prompt and not too much explanation. If you give too many examples or coaching, participants' drawings will not authentically represent their own experiences or ideas.
7. Ask that the participants think aloud as they draw. You can take notes during this time.
8. Once they say they have completed the first mapping, provide the second prompt. Continue this process until you are through all of your prompts.
9. About five prompts is usually enough. Too many more prompts often lead participants to get tired.
10. Once you have the mappings, ask participants questions about what they included. These questions may be aimed at one sketch, or they may tap into how sketches are related to, or different from, one another.
11. Thank the participants for their time.
12. Analyze the sketches, annotations, and your notes. Take note of what was included and what was left out. If using an Ideate method, critically interpret why certain tools and functionalities are desired. How would they change the experience? What are they really providing that your product currently is not? Compare these trends within and between identities, including age, gender, and disability status.

Discussion Questions

1. Have you ever drawn some type of image or chart instead of just taking notes to help you learn about a topic? How did it provide a different perspective than traditional note-taking? If you haven't, can you think of a time it would have been helpful? How so?
2. How can you imagine you could use cognitive maps when your participants express varied ways of processing and communicating information?
3. What do you think you would do if one of your participants told you that they are "bad at drawing?"
4. How is cognitive mapping similar to, and different from, emotional journey mapping?

Notes

1. "The UDL Guidelines," *CAST*, 2018, https://udlguidelines.cast.org/.
2. Edward C. Tolman, "Cognitive Maps in Rats and Men," *Psychological Review* 55, no. 4 (1948), https://doi.org/10.1037/h0061626.
3. e.g., Colin Eden, "Cognitive Mapping," *European Journal of Operational Research* 36, no. 1 (1988).
4. e.g., Daniel R. Montello, "Cognitive Map-Design Research in the Twentieth Century: Theoretical and Empirical Approaches," *Cartography and Geographic Information Science* 29, no. 3 (2002).
5. Michael A. DeVito, Ashley Marie Walker, and Jeremy Birnholtz, "'Too Gay for Facebook:' Presenting LGBTQ+ Identity Throughout the Personal Social Media Ecosystem," *Proceedings of the ACM on Human-Computer Interaction* 2, no. CSCW (2018).
6. Ibid.

16 Brain Writing and Darkside Writing

Even with piles of data from participants in the early empathize stage of the Design Thinking process, UX researchers and designers often find themselves stuck in ruts—tied to old ways of thinking about a platform and doing similar things over and over that rarely lead to long-term, more enjoyable experiences. Brain writing and darkside writing help to think outside the box, offering simple but effective methods that lead to creative solutions to UX problems. While brain writing is more often used with external participants, darkside writing is a great internal method.

Quick Tips	
Tools	Notepad & sticky notes; Google Sheets & Zoom/Skype
Use When	You want to generate new ideas or solutions. You need to brainstorm or organize your thoughts around a research project, or as a research activity in a focus group or interview to generate ideas from participants.
Design Thinking Stage	Ideate

The Methods

Both brain writing and darkside writing are similar to the traditional method of focus groups—they rely on social dynamics to more deeply dive into the issue at hand. Group brainstorming is a well-known process for generating ideas, but it has some issues, including needing a skilled leader and frequent disruptions, especially if there is conflict or if one participant speaks more than the others. But, brain writing allows for a different atmosphere.[1] Studies have found that brain writing sessions generate higher quality ideas than brainstorming sessions.[2]

When implementing brain writing, you ask participants to tackle your Problem Statement head-on. In a group setting, participants are provided with the

DOI: 10.4324/9781003181750-19

current Problem Statement as a simple statement. Each participant begins with a sheet of paper and writes down a solution to the problem. Then, each participant passes the paper to the left. Each time they receive a new sheet, participants build on the ideas that were written previously. The goal is to create as many "ideas" as there are participants. But, instead of creating siloed ideas from each person, you are instead getting a myriad of collaborative ideas that, most likely, no one person would have constructed alone.

For example, let's say you work for Netflix and your Problem Statement is: "Users perceive Netflix to overly promote their original content." One participant may begin by writing a broad, unmanageable solution. But, by the end of the session that one sheet of paper may look like the example in Figure 16.1, held in a Google Doc. You can see how much the solution evolved during the session, providing a great starting point for you as a researcher to make decisions regarding the next step of the Design Thinking process: Prototype.

As mentioned earlier, brain writing is based on the popular method of brainstorming, but, as the researcher, you ask your participants to write down their ideas instead of verbally communicating them. This can really benefit participants who are not as comfortable speaking in front of a group or worrying that they will be ostracized or judged for having a certain opinion or idea. It also helps with those who may talk too much and not allow space for the rest of the group to add their ideas. Because everyone gets one piece of paper, writes down an idea, passes it, and adds to the next sheet of paper, everyone has an equal chance to contribute.

Netflix probably has some sort of algorithm in place to suggest videos to the users, like YouTube. Maybe we change that algorithm.

Obviously Netflix wants to promote their own content, so the recommendation algorithm wouldn't just stop advertising Netflix content. But, maybe there could be better organization for like who sees what when.

I feel like their content is for younger people. What happened to the older shows that we all loved? Why isn't that promoted more?

I think all content is promoted, and I do think that what is shown is related to who is watching. But also they do promote their own content way more and it's just not usually as good.

Maybe there is a way to promote both contents at the same time? Like what if old and new content were presented next to each other? 50/50 split? This way it would be giving equal time, Netflix still gets to advertise their stuff, but we still also see other stuff.

Even better: promote similar content side-by-side. If there is a remake or a show that is obviously trying to re-create an older idea or just similar shows, advertise them side-by-side. This way Netflix still gets to promote their original content, but they connect back to older shows that people loved. This could spark interest in the newer shows too.

Figure 16.1 A sample brain writing session, held in Google Docs.

In darkside writing, another Ideate method, researchers take the Problem Statement and flip it into its opposite. For example, let's say that a team of UX researchers at Facebook are working on ensuring that users trust that Facebook is actively working to fight fake news. Instead of researchers brainstorming how to solve this problem, employing the darkside writing method, they flip the problem, and work together to instead "solve" how Facebook can promote fake news as much as possible. Each member of the team writes at least one "solution" to this new "problem" on a sticky note and sticks it to one side of a whiteboard. Then, each team member takes someone else's sticky and flips the "solution" to actually work toward solving the real Problem Statement.

It may look something like this. One darkside "solution" may be "creating short videos that help users create fake news posts." But, another team member flips it to turn it into a solution for the actual Problem Statement: "creating short videos that help users notice when content may not be reliable." Or, another darkside "solution" may be "edit News Feed sorting algorithms to place suspected fake news posts at the top" that gets flipped into "edit News Feed algorithms to push suspected fake news posts toward the bottom."

You'll want to be careful that flipping darkside "solutions" doesn't become too simplistic. For example, if the darkside "solution" is "edit News Feed sorting algorithms to place suspected fake news posts at the top," the flip shouldn't be "don't edit News Feed sorting algorithms to place suspected fake news posts at the top." Instead, you want to be sure that your team understands that they should be getting creative and providing fresh, new perspectives. The best way to ensure a smooth darkside writing process is to provide examples from unrelated apps or websites that still illustrate how the method should go.

In academic research, brain writing and darkside can be used at any point of the research process when you need to brainstorm or organize your thoughts. These methods can be very useful if you're stuck analyzing your findings and making sense of what your participants said or what you observed. Brain writing and darkside can also be used as research activities in a focus group or interview to generate ideas from participants related to a particular Research Question. They are particularly good activities to understand group dynamics in small group or interpersonal communication.

Use When

Brain writing is best used when you want an external, Ideate phase method. Brain writing is valuable because it places users (or potential users) together in a focus group setting and allows them to collaborate on several ideas that are direct "solutions" for your Problem Statement. As with traditional focus groups, your role as researcher is not to be an interviewer, but a moderator. So, you'll want to use this method if you are comfortable providing the groups with the initial Problem Statement and then letting them work together to create a few solutions. Brain writing is an excellent method for co-creation, a popular concept in the UX industry that means that internal teams and users or customers interpret research findings, generate ideas, and design solutions together.

Darkside writing is best used when you want to conduct an internal, Ideate phase method. Sometimes it makes the most sense for your Ideate process to be internal, maybe due to resource availability: Do you have time for external participants? Do you have a way to reimburse them for their time? Other times an internal Ideate process is necessary just due to the nature of the Problem Statement. It may be that Empathize and Define method results presented a Problem Statement that needs to first be thought through by those who are within the company. Remember, the Design Thinking process is not meant to be linear; you may find at first that you need to conduct an internal Ideate study and then, later, need to conduct another Ideate, but this time using an external method.

What You Need

If working in-person, you only need a few things to conduct a brain writing or darkside writing session. For brain writing, you need pieces of paper and pens and pencils. For darkside writing, you will need sticky notes, pens and pencils, and a space to stick the notes (like a whiteboard or large table).

If conducting the studies virtually, you will want to be sure to use a tool that allows for a strong social, focus group vibe. Services like Zoom work well for participants to chat while you moderate. For brain writing, you should create as many Google Docs as you have participants in each section. Ask each participant to begin in a different Doc, and then rotate to the next, as if they are passing the sheet. Numbering the Docs helps make this process more streamlined.

For darkside writing, we have found a Google Sheet works best. Each participant writes their darkside "solution" in column A. And then, when they are ready, they choose a "solution" and write their flipped solution in column B. While we do highly recommend that these sessions happen synchronously, with everyone on Zoom together, using Google Sheets for darkside writing also helps to create an effective asynchronous darkside session. You can ask that participants visit the sheet at least twice in a certain time period—once to add their "solution" and at least once more to flip another participant's "solution."

Analyzing the Results

Completed brain writing sessions output multiple solutions for your Problem Statement that have been built collaboratively. Completed darkside sessions output lists of one-sentence solutions. In both cases, you will want to analyze them based on a few criteria. Read through the solutions to see which seem most practical. Practicality includes analyzing participants' ideas for fit, possibility, and resources needed. Fitness includes how in line each solution is with your company's goals. Obviously, solutions that propose huge overhauls or changes that would completely reshape the purpose of your app or website are not practical. Similarly, keep in mind what is possible from a programming and design perspective. Some changes are just not possible because they sway too far from what the app or website is already designed to do. Finally, try to make a list of what resources are needed to complete each solution. Would more research be needed beyond the testing phase? How many people would need to be involved? How much time is needed? Are there deadlines that need to be met or a budget to keep? Is your team large enough to complete it? Are you working alone?

Case Study

In 2018, researchers from Denmark conducted research in an effort to create a mobile app that teens with cancer could use to communicate and, hopefully, experience an increase in health-related quality of life. The researchers note that, while there are apps for teens with cancer, these existing apps were not created with data from the actual teens themselves. Thus, their goal was to use co-creative methods to suggest a more inclusive and effective user experience.[3]

One method employed by the researchers was brain writing. Using this method, they found that their participants would like an app that includes a community forum, an information library, and a symptom and side-effect tracking tool. The researchers also learned that bright, warm colors are preferred. This case study is a great example of how we need to remember that we are *not* the users. Previous apps were created by people that were quite unlikely to be the actual end users. But, with a little insight from teens, through a UX method like brain writing, the researchers were able to begin designing a product that could have real impact on the lives of many young adults.[4]

Steps

1. Organize your participants.
    a. If working in-person, find a space that will work best for brain writing or darkside writing.
    b. If working virtually, set up the Zoom meeting and either the Google Docs or the Google sheet.
2. Provide whatever tools and instructions are necessary.
3. Obtain any demographic information you may from your participants. This may be completed via a short survey or by just asking them a few quick questions before the writing session begins.

4. Define the Problem Statement. Remember that it may not be exactly as you have defined it internally. You may need to change convoluted language or jargon.

    a. If conducting brain writing sessions, remember you will essentially be providing your Problem Statement, perhaps with some modifications, to your participants.
    b. If conducting darkside writing sessions, you will present the "flipped," "dark" version of your Problem Statement.

5. Begin the session.

    a. For brain writing, ask that each participant takes a set amount of time to write a one-two sentence solution to the Problem Statement.
    b. If conducting a darkside writing session, ask that each person takes a set amount of time to create a "solution."

6. After the set amount of time has lapsed, move on to the next step.

    a. For brain writing, ask each participant to pass their sheet of paper to the right. If you are working virtually, you can ask that each participant moves to the next number sheet. For example, if you have eight participants, you will label the Docs 1–8. A participant that began on Doc 2 would move to Doc 3. The participant on Doc 8 would move to Doc 1.
    b. For darkside writing, ask participants to take a sticky from the board or table that they did not write, read it over, and create a new sticky that flips that dark "solution" into a solution for your actual Problem Statement.

7. Repeat these steps until the session is complete.
8. Thank your participants for their time.
9. Begin analyzing your solutions. Remember to pay attention to fit, possibility, and resources needed. Also remember to consider demographic and cultural differences.

Discussion Questions

1. Think of two Problem Statements or Research Questions, one that would call for a brain writing session and one that would call for a darkside writing session. Why do you think each would be best for their related method?
2. Why do you think the social dynamic of focus groups is useful in UX research if most device usage is usually a solo, personal activity?
3. How could you get creative with these two methods and adapt them for participants who find it difficult to find their voices through writing?

Notes

1. Arthur B. VanGundy, "Brain Writing for New Product Ideas: An Alternative to Brain-storming," *Journal of Consumer Marketing* 1, no. 2 (1984), https://doi.org/10.1108/eb008097.
2. Peter A. Heslin, "Better Than Brainstorming? Potential Contextual Boundary Conditions to Brain Writing for Idea Generation in Organizations," *Journal of Occupational and Organizational Psychology* 82, no. 1 (2009).
3. Abbey Elsbernd et al., "Using Cocreation in the Process of Designing a Smartphone App for Adolescents and Young Adults With Cancer: Prototype Development Study," *JMIR Formative Research* 2, no. 2 (2018), http://formative.jmir.org/2018/2/e23/.
4. Ibid.

17 Card Sorting (and Tree Testing)

When you think about an app or a website, you may think about its home page or maybe you think about a few of the specific features you like about it. What you may not think about are all of the different pages, options, tools, menus, and so on of a specific platform. This is not accidental. Digital spaces don't want their users to notice clunky menu options or to get lost moving through pages. Instead, they want users to easily find what they are looking for; they want to provide a logical, easy-to-follow organization. Here is where card sorting comes in. Card sorting allows participants to take the role of designer, organizing content on a website or an app in a way that makes sense for them.

Quick Tips	
Tools	Index cards or Optimal Sort at optimalworkshop.com
Use When	You need to discover how to best organize page structure or menus or categorize content.
	Your research relates to how people categorize information around a particular topic or in a particular context or where people go to find specific information.
Design Thinking Stage	Ideate

The Method

Card sorting allows users to take a hands-on approach to helping designers organize content that is presented in digital products. As the researcher, you are essentially asking participants to become a part of your app or website's organization design. Card sorting is used often because it is quick, cheap, and reliable.[1] Card sorting is particularly useful when your Problem Statement involves findability or usability or when your Research Question involves understanding how people categorize information.

Card sorting comes out of the "Q Methodology," created by psychologist William Stephenson. Participants perform "Q sorts," where they rank

DOI: 10.4324/9781003181750-20

statements written on cards based on some provided context so that researchers can better understand their viewpoints on a particular topic. Cards can also be images or videos (if the sorting is completed on a digital device).[2]

There are generally two types of card sorting—closed sort and open sort. In both scenarios, you provide participants with cards that have the content you want to be organized on them. For example, let's say you are helping a new department store provide a better user experience on their website. Each card would list a different item that the store sells. You may not be able to include *every* single item, because that might get a little tiring for your participants. But, including a representative sample (e.g., sneakers, sandals, flats, dress shoes, etc. as part of the "shoe" inventory) will still work. In a closed sort, you first provide the participants with the different categories the store wants to include—accessories, teen clothing, baby clothes, shoes, kitchen appliances, furniture, and so on. Then, participants match up the item cards with what they believe to be the appropriate category. In an open sort, participants are given the item cards, but the category headers are blank; it is up to the participants to decide the categories where the different items fit in. Card sorting is often used to categorize menu content on a webpage, for example, what sort of stuff should go on the "about" page, what fits under the "FAQs," etc.

You could also design some hybrid of the two types of card sort. Maybe you provide the participants with some categories but also provide additional blank cards so they can add more categories that you didn't think of. Maybe there are some blank item (content) cards. Or maybe participants have the ability to edit created cards. If you have the time and resources, you can also do multiple phases of card sorting. For instance, you could start with an open sort to allow your participants to create categories that they think the different items would fit under. Then, in phase two you could conduct a closed sort to test whether the categories that came out of the initial open sort work for other participants. Any combination can work, as long as you match what you want to learn with what you are asking your participants to do.

At this point, it's important to also briefly cover tree testing, which is basically the reverse of card sorting. It is a usability technique (often presented as part of usability testing) that helps researchers validate whether or not the information architecture (how content is laid out, organized, and categorized on a website or app) makes sense. Tree testing is called "tree" testing because the information architecture of a website typically looks like a tree, with a core trunk (the home page) and large branches with smaller and smaller branches shooting off it. (Tree testing is not to be confused with problem trees, as discussed in Chapter 14!)

Whereas in card sorting, participants are first given a set of cards and asked to categorize them, in tree testing, they are provided a final text version of a site map (basically all the menu categories of a website and all their submenus) and asked to highlight where they think a particular piece of information or items would be found in this categorization.

Card sorting comes before tree testing; tree testing is essentially a way of confirming the results of a card sort and validating the general findability of your site. As you can probably guess from the name, tree testing is commonly used during the Test stage. An example of a tree test would be to provide a participant with a site map of a shopping website and ask them to find their "cart" or to find information on how to return an item. Following the user's reasoning while they navigate, the researcher can get some ideas on what improvements need to be made to make the site more easily navigable and the information (e.g., carts or return policies) on it more findable.

Tree testing provides qualitative data in terms of reasoning by the participants (of course, this is only possible in a moderated tree test where participants are asked to talk out loud as they navigate), but it can also provide quantitative data, such as, how many participants correctly completed the task and how long did it take them? Tree testing is typically done with more participants than other qualitative methods, because of the quantitative component that needs more participants in order to be seen as valid.

Use When

Card sorting is mostly used during the Ideate phase of the Design Thinking process. Remember, in the Ideate phase, the goal is to think outside the box and begin exploring data-driven solutions to your research problem. Often, methods within the Ideate phase are internal—you or you and your team (inside the company) work together, looking at participant data and brainstorming changes to a process or interface design. Card sorting is unique because it brings participants—actual users—into the Ideate process.

The card sorting method is best used when you already have an idea of what a space will include but you aren't completely sure how those things should be displayed, organized, or grouped. For example, let's say you are helping to get an exercise app up and running. The developers know that they want to include different types of workouts, offering slightly different workouts depending on equipment needed and time available. But they aren't sure how to best group these varied workouts. The card sort method could take workout type, length, and equipment needed and ask participants to organize them into what they view as relevant and intuitive.

Card sorting can be used as a creative method for any academic research that is interested in understanding how people categorize information. Qualitative academic research is typically interested in understanding what, why, and how. So, for example, you could ask different people to categorize workouts (as in the example provided earlier) and then ask them why they categorized them as such. It might be interesting to see whether different groups categorize workouts differently and what the reasoning for this is. For instance, you might find that participants of varying genders categorize the same type of workout (e.g., yoga) under different categories (e.g., strength training? cool-down?). You could interpret this as evidence that gender stereotypes around exercise (and what is perceived as a "real" workout) persist.

What You Need

In person, card sorting is as easy as getting index cards and writing a piece of content (an item) or a category on each. You then ask participants to move these cards around on a big table or workspace. This can also be done in groups or individually to get different perspectives, similar to how individual interviews output different findings to focus groups—the group dynamic builds on itself, providing insights on group interactions, rather than singular attitudinal data.

Virtually, websites like OptimalWorkshop.com provide a tool—Optimal Sort—that allows participants to use their own computers at home to drag and drop cards to different areas on a screen, similarly to how they would move index cards around on a table. Beyond just sending a link to the Optimal Sort workspace participants, you can bring participants together using software like Zoom, Google Hangouts, and Skype and have them work on the virtual card sort together. In the in-person model, be sure you have an efficient way to take note of the different ways participants organize the cards. You can choose to take notes, but a better way is to take a photo or record the interaction. Online, Optimal Sort not only saves the structures for you but provides some initial analyses like similarity matrices and dendrograms (taxonomic, tree diagrams).

In both in-person and virtual card sorting, it is important to randomize the order in which you present participants with the cards to prevent order bias. This is commonly done by "shuffling" the stack of content cards that you are asking participants to sort. You wouldn't want your findings to be driven by participants unconsciously thinking that the first few cards are the most important or by participants beginning to feel tired by the end and not thinking as deeply about the same few final cards. It is up to you if the category cards remain in the same order for each participant. If you already know how the labels will be ordered on your site, you can present them in that order for each participant. If you aren't sure yet, or if you don't want the category cards' order to play a role in your findings, mix them around. In-person, randomizing cards to prevent order bias involves you shuffling the cards before each new participant comes in. Online sites like Optimal Sort provide a checkbox, and, once ticked, the cards shuffle automatically for each participant.

Analyzing the Results

Perhaps surprisingly, the exact organization of a space is not really what we are looking for when analyzing card sorting results. It is unlikely that any two participants are going to create the same exact structure, and so, as the researcher, you will find your insights by looking deeper than just paying attention to how each participant organized the cards. The conversations you have with participants and what you learned from hearing them think out loud and explain their choices are really the most beneficial pieces of the collected data. To analyze your results, you take your sorted cards and comments from participants and tease out common themes.

Of course, you can look for general trends regarding pieces of content that participants seem to frequently place under the same category. But, also take note of why these choices were made. What felt obvious to participants? What seemed a little more specious? Why did they make these decisions? Were they based on personal experiences? What other similar digital spaces do they frequent that could have influenced these results? What are some common trends or standards that users expect to see across apps or websites? Your sorted findings may be more about what the participants are used to than about what actually makes sense!

It is important to remember that asking someone how content should be sorted is not the same thing as asking them what they think of the content or which pieces of content they would use or frequently view. When participants are provided with 50 cards, they are most likely going to attempt to sort all of those cards under your defined categories or categories of their own. But, don't assume that this means that every user will like and use all 50 of those things. Remember it is important to consider your study in context. Do not use card sorting as a method to gain user perceptions or opinions of *content*. Instead, keep in mind that you are simply investigating usability and findability through *organization schemes* that users believe to be most intuitive at that moment.

Don't expect to get an exact model for organizing your website or app (or in academia, for people to fit into neat groups that organize content exactly the same way). It is unlikely that you will get exactly the same content organization from your participants. The goal, instead, is to look at the choices participants made and *why* they made them. Pay attention to who they are, why they use the space, and what is important to them. These will help you in identifying why the space makes sense to them with the structure they laid out. Then, it is up to you to take a critical eye and decide on the best suggestions for page and menu organization. Choices you make, including menu and submenu order or what is more/less visible, play an important part in what you communicate to users about the importance, norms, and expected uses of a space. These information architecture choices can then be designed via rapid prototyping (usually by a UX designer or by you) and tested.

Case Study

Alison Doubleday, a researcher at the University of Illinois at Chicago, used card sorting to study students' experiences with a new science curriculum on the course management site Blackboard. To begin, Doubleday conducted an open card sort, before the new curriculum was implemented. Cards were created by faculty, and card names were based on the content that was previously available on Blackboard. The participants were asked to group these cards and then to name the groupings. Groupings included Course Information (with content like syllabus, course schedule, and course policies), Educational Resources (with content like research articles, animations, and links to websites), and Daily Materials (with content like online lessons, tutorials, and case studies). After analyzing the results, the Blackboard site for the new course was created.[3]

Doubleday then conducted another card sorting study to see what, if any, changes were needed to make the Blackboard site more user-friendly. But this time, the research was a semi-closed sort. The same cards were used from the open sort, but the categories Doubleday decided on after the open sort were given to the new participants (instead of letting these participants create their own groupings). The participants were asked to move the cards under these categories, but they were also allowed to edit group names, delete group names, and add new group names.[4]

Doubleday's results from the second semi-closed sort were actually slightly different than the first. This points both to the importance of understanding how different types of card sorts can provide you with different results and to the importance of repeating a Design Thinking stage if necessary. As with much UX research, the card sorting results were surprising to the research team. Had they used anecdotal evidence, they would have organized the site completely differently. Remember, you are not the user—you need to do research to understand how users *actually* think and behave!

Steps

1. Decide how you will conduct the study. Choose if you will do an open sort, a closed sort, or some combination of the two. Remember to let your Problem Statement or Research Question guide your decision.
2. Prepare the study materials. Card labels should be short enough that participants can read them quickly, but also descriptive enough that they are clear.

   a. If you are conducting the card sort in-person, create your index cards for sorting.
   b. If you are conducting the study virtually, create the digital cards using a site like Optimal Workshop.

3. Conduct the study.

   a. If working in-person, bring participants in one at a time, explain their task to them, and watch as they sort the cards.
   b. If working virtually, send the link to your participants. You can choose to have them do the sort on their own time and look at all of the results after. Or, you can choose to conduct a synchronous card sort, using a service like Zoom to watch the participants as they sort the cards, so the process is closer to an in-person session.

4. After the cards are sorted, ask participants (if they are in-person or meeting synchronously online) questions based on the choices they made. Find out why they made the choices they did.

5. Be sure to save the information architecture (the classifications and organization) that your participants create. If your study is virtual, tools like Optimal Sort save your results for you. But, if you are working in-person with index cards, be sure to take an overhead photo of each structure or to record the session. Save these images with your notes so that you can organize them together.

6. Analyze your data. Remember that you are not looking for some exact structure or for every participant to provide the same exact organization. Instead, you want to study the structures created, as well as your notes from asking questions of your participants. Understand the choices made and why those choices were made. Then use these, along with your expertise and findings from your Define stage, to move into the Prototype phase, and create a design with the most intuitive organization structure.

Discussion Questions

1. How would you study your college's online course management system (Blackboard, Canvas, D2L, etc.)? Try creating cards out of the content contained in a specific course site. How would you organize an open sort study? A closed sort? Something in between?

2. Can you think of an example when card sorting would work better as images or videos instead of words?

3. How could you still capture participants' thoughts if they complete a card sort virtually and asynchronously (without you watching)?

Notes

1. Donna Spencer and Todd Warfel, "Card Sorting: A Definitive Guide," *Boxes and Arrows*, April 7, 2004, para 4, https://boxesandarrows.com/card-sorting-a-definitive-guide/.
2. William Stephenson, *The Study of Behavior: Q-Technique and Its Methodology* (Chicago: University of Chicago Press, 1953).
3. Alison Doubleday, "Use of Card Sorting for Online Course Site Organization Within an Integrated Science Curriculum," *Journal of Usability Studies* 8, no. 2 (2013).
4. Ibid.

18 Heuristic Evaluations and Critical Analysis Walkthroughs

A walkthrough is a usability method where the researcher themselves or a usability expert navigates through the website or app with the goal of evaluating it based on some predetermined criteria. Thus, in the walkthrough method, there are no participants in the traditional sense, and instead, the researcher or expert role-plays the user. This method allows researchers to understand first-hand how a person might navigate through a website or an app. In this chapter, we hone in on two specific walkthrough methods: heuristic evaluations and critical analysis walkthroughs (used in academia). (Note that there are many other types of walkthroughs, including the very popular cognitive walkthrough, where a novice navigates through the website or app to test its navigability and usability.)

A huge advantage of walkthroughs is that they are cheap and easy to conduct, as they do not require participants or any special equipment. The person doing the walkthrough and the website or app itself are the only things necessary for this method, though you can choose to record a walkthrough for note-taking purposes, in which case you also want recording equipment.

Quick Tips	
Tools	App/website to test Screen recording software or video camera
Use When	You want to understand the usability issues in an app or website. Your research involves evaluating how bias and assumptions are baked into the app and how digital spaces guide user behaviors in non-neutral ways.
Design Thinking Stage	Test

The Method

In a walkthrough, the researcher or other assessor downloads the app or accesses the website and moves through it like a typical user, signing up,

DOI: 10.4324/9781003181750-21

navigating through different screens, and eventually logging out. As they move through it, they will note their thoughts and especially note parts of the process that are difficult or not intuitive.

A heuristic evaluation or expert review is a specific type of UX walkthrough conducted by a usability expert (who could be the researcher themselves, if they are a usability expert!), to check that the website or app meets certain standards of usability. Walkthrough methods help the researcher see where there are usability problems with the site, but they're also great for making explicit the implicit values and biases present in apps and websites, from an academic research perspective. Ben Light, Jean Burgess, and Stefanie DuGuay propose a walkthrough method for academic researchers as a way of conducting a critical analysis of an app, bringing to light certain assumptions that the app is based on, and pointing out the limitations resulting from these assumptions.[1] For readability purposes, to distinguish this method from other walkthrough methods, we have termed this a "Critical Analysis Walkthrough" in this chapter.

Walkthrough: Heuristic Evaluation

A heuristic evaluation, also known as an Expert Review, is a type of walkthrough method where a usability expert checks over a website to make sure that it is user-friendly and to uncover any issues with the navigation and interaction with the site. A heuristic evaluation can be done at any stage of the design process (remember, design thinking is an iterative process!) but is typically done in the testing stage.[23]

A heuristic is a rule of thumb, principle, or standard to evaluate something on—in the case of UX, the usability of an app or a website. There are hundreds of different heuristics that UX researchers can use to evaluate websites or apps, but the "best practice" guidelines in the field are Jacob Nielsen's "10 Usability Heuristics for User Interface Design":

1. Visibility of system status: When users interact with the website or app, they should always know what is going on in the system, through immediate feedback on every interaction between the user and the system. For example, when you click "download" on a website, you will get a pop-up window telling you the download has started and keeping you informed as to the percentage complete as the file is downloading.
2. Match between system and the real world: The language and concepts used on the website or app should be natural to the user and free of jargon. Use terms that users would typically use in their daily lives, not technical phrases and concepts. For example, most software uses the icon of a trash can as a space designated for deleted items.
3. User control and freedom: Users should be able to move through the system backward and forward unhampered, and have the option to go back

if they make mistakes or change their minds. The options to "undo" and "redo" should be part of every system.

4. Consistency and standards: Every website or app should follow platform conventions, such as using the "home" button for the landing page or being able to scroll down a page through the scroll bar on the right-hand side. Check out Apple's Human Interface Guidelines[4] and Google's Material Design Guidelines[5] for consistent design guidelines.

5. Error prevention: The system should guide users through it in a way that makes errors less likely, through built-in constraints (e.g., in a phone number field, the option to add letters should be disabled, "forcing" the user to type in numbers) and confirmation messages asking a user whether they definitely want to commit to an action.

6. Recognition rather than recall: Users should be able to move through the system easily without having to remember parts of the system. That is, all possible options should be clearly visible or easily retrievable at all times (think of the menu bar in Microsoft Word and how it presents the user with a range of options, such as save, edit, etc. using intuitive icons and text).

7. Flexibility and efficiency of use: The design should allow both experienced and novice users the ability to move through the system efficiently. For example, keyboard shortcuts allow seasoned users to quickly produce certain frequent actions. Flexibility, in the form of customization and personalization based on personal preferences, should also be part of the system.

8. Aesthetic and minimalist design: The design should contain only what is necessary for efficient and enjoyable use. Cut out unnecessary words, decorative but distracting features, and other noise from the system design.

9. Help users recognize, diagnose, and recover from errors: Error messages should be precise, should not use codes or jargon (rather, use plain, real-world language), and should offer a solution to the problem.

10. Help and documentation: Help and documentation information should be easy to find and concrete in the solutions presented.[6]

If possible, a heuristic evaluation should be conducted by more than one person, where each review is conducted separately and then the review data are synthesized. In this way, you're more likely to catch errors or issues that just one assessor might have missed (having more sets of eyes during any review process is always a good thing). Typically, you should aim for three to five reviewers if the budget allows (having more than three to five experts yields diminishing returns, in that they find fewer unique problems than only three to five reviewers would). If your budget does not allow for the hiring of usability experts, you can conduct an informal heuristic review yourself as the researcher, by walking through the site methodically (from start to end, as a user would) while carefully considering your predetermined heuristics.

Walkthroughs: Critical Analysis Walkthrough

Light, Burgess, and DuGuay outline how a walkthrough can be used to critically analyze an app in an academic context. In this case, the researcher first delves into the app's background information, such as its mission and vision, how the app makes money, and how it is governed (who sets the policies and rules for use and how are these implemented). Next, the researcher moves through the app systematically (signing up, using it as a typical user would on a daily basis, and eventually logging out/discontinuing use) while paying close attention to the app's intended purpose and any assumptions around its ideal use that are present in the design. During this "technical walkthrough," a researcher will interact with the app as a typical user would, tapping buttons, exploring menus, and reading through screens.[7]

The main point of this type of walkthrough is to try to understand how an app guides a user through it in specific ways and what does that mean more broadly about cultural or societal values that are embedded in the app design—and what it practically means for certain groups of users in terms of accessibility and inclusivity.

So, the big difference between a heuristic evaluation in UX and a critical analysis walkthrough in academia is the *purpose* of it and what the researcher pays attention to during the walkthrough. In conducting heuristics evaluations, the researcher aims to learn about usability issues (what challenges might a user face in navigation and whether the product meets certain standards to make it user-friendly), while in the critical analysis walkthrough, the researcher aims to learn about how different societal and cultural values shape app design (and what this means for the users, often in the sense of biased or exclusionary experiences).

Use When

Walkthroughs are used in the Test stage of the Design Thinking process, to ensure a website or an app works the way it should or to check through what improvements could be made to an existing product. In a heuristic evaluation, the website or app is checked against a list of standards that ensure the product meets them.

In academia, researchers can conduct critical analysis walkthroughs to better understand how an app or a website guides the user through it in particular ways that are connected to broader societal norms and expectations. In this type of walkthrough, researchers can use features of the app as evidence to show how the way the space is designed connects to cultural biases and assumptions. A researcher will navigate through the app and, for instance, find that at sign-up, the app provides limited gender options, forcing the user to pick between male and female only. This assumption about the gender binary presents limitations to users who do not identify as male or female.

Both heuristic evaluations and critical analysis walkthroughs can be combined with user research, such as interviews or usability testing, to include the perspective and actual behaviors of the users in the analysis.

What You Need

Walkthroughs do not need any special equipment—all you need is the website or app to be evaluated and a way of taking notes. You (or the assessor, if you're hiring a usability expert) can either think out loud as you walk through and record yourself (it's helpful to record the screen as you walk through so you can clearly see what you're referring to) or you can jot down your notes. You might consider taking screenshots of the walkthrough as you're going if you're not able to video record, as it's easier to refer to pictures for design purposes. (Because a walkthrough does not involve participants, the consideration of in-person versus remote walkthroughs is moot.)

Analyzing the Results

To analyze the results of a heuristic evaluation, the expert (or experts if you have more than one) will go through their notes and create a list of usability problems connected to the principles or standards that were used during the evaluation. You as the researcher will then take these lists of usability problems and compare across experts (again, if you have more than one). Are there some issues that all the experts agree on? If yes, these should be prioritized as fixes in the next iteration of your design. If you don't have more than one expert, you still need to prioritize the usability problems list based on importance and feasibility. You need to consider whether you have the time, budget, and other resources to actually do the improvements. Sometimes the experts themselves can provide some advice on fixes or redesigns along with the list of usability issues.

In critical analysis walkthroughs, as the researcher navigates through the app, they will be noting down actual observations (e.g., the dropdown menu lists items in a particular order) and they will also be considering what this means (so doing an analysis as they go), connecting their observations to previous research and knowledge of cultural/societal norms and expectations (e.g., the dropdown gender selection lists male first, then female, then non-binary—this could be interpreted as ranking these identifications in order of societal importance or acceptance). As the researcher is doing the walk-through, they should be doing so with a critical eye. What actions does the app guide the user to do? How does the interface constrain what the user can do? What does this mean about expectations of use and the "ideal" user behavior? Are there biases or assumptions baked into the design in a way that privileges some users over others? A critical analysis walkthrough is typically complicated and nuanced, so for a more detailed explanation see the references section.

Steps: Heuristic Evaluation

1. Decide on the heuristics to be used for evaluation.
2. Each expert then separately conducts a review based on the predetermined heuristics. The expert should go through the system at least twice, once to

get a general sense of the system and once paying attention to any usability issues. Such a review typically lasts one to two hours.

    a.   Reviews can be recorded either as written comments in a document or as a verbal/think-aloud walkthrough of the site.

3.   The review output should be a list of usability issues in the system, each connected to the heuristic they don't adhere to.
4.   The lists of usability issues are then combined and consolidated to rank the most pressing issues within the website or app, to guide design changes.

Steps: Critical Analysis Walkthrough

1.   Map out the "environment of expected use" by finding information on the app's:

    a.   Vision—what is the app's purpose and target audience?
    b.   Operating model (business strategy and how the app makes money)
    c.   Modes of governance (rules and regulations around use, e.g., Terms of Service)

2.   Conduct a technical step-by-step walkthrough. During this process, you will explore all the buttons, menus, and screens of the app, paying attention to things like how the UI is arranged, what are the different available functions and features, the branding and colors, etc. The technical walkthrough should include navigating the app through:

    a.   Registration and entry
    b.   Everyday use
    c.   Logging out/discontinuation

3.   Analyze your findings with a critical eye, paying attention to biases and assumptions baked into the design.

Case Study

A great example of all the steps in a heuristic evaluation can be found in an evaluation case study of yuppiechef.com, a South African-based online kitchenware store, conducted by UX specialist Marli Ritter. As a heuristic evaluation is systematic and very long (the assessor must do a thorough job of assessing the *whole* site or app on *multiple* criteria or heuristics), we provide some highlights here but encourage you to read the entire case study.[8]

One of Nielsen's ten heuristics is "match between the system and the real world." Yuppiechef.com failed this standard in one section of the site, by using confusing (not standardized) icons for the action of repeat subscriptions of kitchen items—a calendar icon when, according to Ritter, "a more suitable icon would be something that illustrates continuous movement." However, in another area of the website, the "match between system and real world" standard was "passed"—images of actual handwritten notes that accompany gifts were shown to illustrate what happens when you mark your order as a "gift." This is a great example of using a real-life object in the digital experience.[9]

As you can see just from this short example, heuristic evaluations can be very complicated because they are holistic, looking at every part of a website (so different parts might pass/fail the same heuristic), and also because they are so detailed. However, a thorough walkthrough is worth it in terms of really making sure that your product is fully user-friendly from start to end!

Discussion Questions

1. Think of some different scenarios for when you might want to conduct a heuristic evaluation and when you might want to conduct usability testing (getting novice or typical users—not experts—to move through an app or website). What are the pros and cons of each?
2. Conduct a walkthrough as a critical analysis of your favorite app. What are some assumptions that the app makes about you as a typical user? Are there parts of the app's design that frustrate or delight you—when you feel it is "not getting you at all" or "really speaking to you"—what are these parts and why do they make you feel that way? If you were to redesign the app to be more "for you," what would you change about it?

Notes

1. Ben Light, Jean Burgess, and Stefanie Duguay, "The Walkthrough Method: An Approach to the Study of Apps," *New Media & Society* 20, no. 3 (2018), https://doi.org/10.1177/1461444816675438.
2. Jakob Nielsen and Rolf Molich, "Heuristic Evaluation of User Interfaces," *Proceedings of the SIGCHI Conference on Human Factors in Computing Systems*, 1990, https://doi.org/10.1145/97243.97281.
3. Jakob Nielsen, "10 Usability Heuristics for User Interface Design," *Nielsen Norman Group*, November 15, 2020, www.nngroup.com/articles/ten-usability-heuristics/#poster.
4. "Human Interface Guidelines," *Apple*, accessed July 18, 2021, https://developer.apple.com/design/human-interface-guidelines/.
5. "Guidelines," *Google*, accessed July 18, 2021, https://material.io/design/guidelines-overview.
6. Nielsen, "10 Usability Heuristics for User Interface Design."
7. Light, Burgess, and Duguay, "The Walkthrough Method."
8. Marli Ritter, "Heuristic Analysis of yuppiechef.com: A UX Case Study," *UX Collective*, October 7, 2018, https://uxdesign.cc/ux-case-study-heuristic-analysis-of-yuppiechef-com-c92098052ce4.
9. Ibid.

19 Usability Testing

Usability testing is a research method of observing users using a website or an app. During usability testing, researchers can see first-hand if users can easily and efficiently complete tasks with the technology in a real-world context. Typically, in a usability testing session, a user will be provided with a set of tasks and will be asked to think aloud as they move through the app or website (e.g., "buy an XL blue shirt" on a store app). In this way, the researcher can see first-hand if there are problems with site navigation or if the design is not intuitive in other ways. Usability testing is done with mock-ups, wireframes, or prototypes, to allow for further fixes before product launch.

Quick Tips	
Tools	Task list Camera or share screen capabilities (if unmoderated testing)
Use When	You want to test a product for usability. Your research is about accessibility and usefulness of communication technologies and devices by people with different abilities, identities, and cultures.
Design Thinking Stage	Test

The Method

Have you ever used a website where you couldn't figure out how to find something simple or where you kept clicking on multiple broken links? Experiences like this are common and can be very frustrating to users. This is why usability tests should be conducted throughout the design process to avoid such scenarios. It's important to make sure that a website or an app is usable and that users can move through it without problems. Areas of focus in usability testing include testing whether users can navigate through the product easily, whether the product is effective (it does what it says it does), whether it can

DOI: 10.4324/9781003181750-22

be navigated efficiently (within a reasonable timeframe), as well as testing for accessibility (people with different abilities can navigate it).

Usability tests consist of three important pieces: the researcher (also known as the facilitator or moderator in this instance), the user, and tasks for the user to complete using the digital product. The researcher provides the tasks for the user to complete on the website or app and then observes the user and asks for feedback as the user moves through the tasks.

Usability tests can be run in-person, with the facilitator and user in the same room looking at the device while the user navigates, or they can be run remotely. In remote usability tests, the user shares their screen with the researcher, either using screen sharing on their computer or using a camera pointed at the screen, as they complete the tasks.

Aside from being categorized as in-person or remote, usability tests can also be categorized as moderated or unmoderated. In moderated tests, the researcher guides the user through a series of tasks and prompts them to talk about what they are doing and why. In an unmoderated test, the user is provided a list of prompts and tasks in written form that they move through on their own. Because unmoderated tests are always recorded, the researcher can see both the user behavior on the website or app (where they click/hover, how they navigate through the website, etc.) and the reasoning behind the behavior. Some unmoderated tests even include a recording of the participant's face, so that the researcher can see nonverbal reactions to the website or app. There are many online services, such as usertesting.com, that recruit participants and run remote usability tests for researchers.

Usability testing can be either qualitative or quantitative, though qualitative testing is more common. And, even more commonly, the two are used together to complement each other. Qualitative usability testing focuses on observations and think-aloud comments to understand the problems in a website or app. Quantitative usability testing typically focuses on two metrics: task success (whether the user completes the task or not) and time-on-task (how long it takes a user to complete a task).

A tool used frequently in quantitative usability testing is the System Usability Scale, which provides a quick measurement of usability based on ten questions asked of the user after going through the task, with answers ranging from "strongly agree" to "strongly disagree." The questions are listed.

1. I think that I would like to use this system frequently.
2. I found the system unnecessarily complex.
3. I thought the system was easy to use.
4. I think that I would need the support of a technical person to be able to use this system.
5. I found the various functions in this system were well integrated.
6. I thought there was too much inconsistency in this system.
7. I would imagine that most people would learn to use this system very quickly.

8. I found the system very cumbersome to use.
9. I felt very confident using the system.
10. I needed to learn a lot of things before I could get going with this system.[1]

Scores from users provide an estimate of the ease of use of a website or app (for scoring, see https://hell.meiert.org/core/pdf/sus.pdf).

As a more qualitative example, let's say you have created a website for selling chairs. You have a large inventory, with hundreds of different products on offer. Importantly, you are also able to provide customization for some chair styles—so customers can choose the material, the size, the leg height, etc., that suits them best. You now want to test whether people coming to your website understand what types of customization are available so they can easily place a unique custom order. You start your moderated usability session by showing participants the home page of your website and you task them with "finding out what customization options are available." The participants will then explore the site, talking out loud as they are navigating.

Most of the participants easily find the "customize" button for select chair styles, navigating there directly with a few clicks. So far, so good. Once they select the "customize" button, they are presented with a list of customization options. You ask the participants to explain to you, from looking at this list, what customization options they think are available to them. A few participants mention at this stage that they're not sure whether the word "color" in the customization options refers to the frame or to the cushioning of a chair. So, when they see the different colors provided at that option, they're not actually sure what the final customized chair will look like. You can now use that finding to change the design of the site and create two separate items in the customization menu—"frame color" and "cushion color." Remember, usability is all about ease of use, which includes findability of content and eliminating frustration or confusion for the user as they move toward their end goal (in this case, buying a chair customized to their liking).

If you want to take it to the next level, eye movement tracking and heat maps are two specific UX techniques that can be used during usability testing to get an even more precise understanding of how users navigate through a digital space, rather than simply watching people navigate and speak their thoughts out loud. As the name suggests, in eye movement tracking, a sensor measures user eye movement to see where a person is looking on a screen. Knowing where a person looks first or where their gaze lingers can tell a researcher where points of interest (or possible trouble spots) are on the page, even if the participant is not actually conscious of this (and might not have told you through thinking out loud). Heat mapping is a behavioral analytics tool for websites (companies such as Hotjar provide heat mapping services). It is similar to eye movement tracking except that software logs where a user moves, hovers, and clicks on a page and then the software presents the researcher with a heatmap of the most used areas. These tools are useful in providing more objective, physiological measures of user experience in conjunction with think-aloud and observational data.

Usability testing is most similar to experimental methods in scientific research, where users are provided with a task in a particular scenario and then the researcher observes them completing this task, to test whether the research hypothesis is supported or refuted (e.g., people who are sleep deprived will perform worse on a reading task than those who are not sleep deprived). The goal of usability testing is to test an entire website or app, so it's a much more holistic method than an experiment testing a specific hypothesis.

Use When

Usability testing helps to uncover problems in a digital product and also provides ideas for opportunities for new features or improvements. It's important to conduct usability testing throughout the product design process because even the best designers can't fully know how actual users will interact with a product in a real-world context.

Usability testing is useful in Media and Communication Studies for answering research questions related to cultural differences, bias, and inaccessibility of certain technologies by people with different abilities, identities, and cultures. Usability testing with these various groups can bring to the surface problems with the ease of use of websites for particular groups, thus providing evidence for pervasive unequal power dynamics built into tech.

What You Need

In order to conduct a usability test, you need a wireframe or prototype of the app or website you want to test. Wireframes and prototypes are iterations of the product throughout the design development. No app or website is fully developed all at once but is, instead, created through various stages, with layout, content, and interactivity being added and adjusted at different points in the product design process.

The difference between wireframes and prototypes is that wireframes are low-fidelity (not fully fleshed out) representations of the final product (such as a web page), while prototypes are high-fidelity representations (they look and feel more like the final product will). Wireframes can be created using pen and paper sketches that users can then "click" through by moving the pieces of paper around. Prototypes are usually digital versions of the website that do not have all the back-end functionality (code, etc.) but are still interactive on the front end.

You also need to create a task list for users to move through ahead of the test. You can either give a simple vague task, such as "buy a blue XL shirt" or provide more detailed instructions and see if users can navigate through the task easily. Just asking a user to explore the site is not enough direction to test usability—a user will likely feel overwhelmed and stuck with such a request. The wording of tasks is really important. You don't want to direct the participant too much in a particular direction (this is known as "priming" and can skew your results). Instead, you want to provide just enough direction that they know where to start but where they still have some autonomy in their navigation to allow you to understand how the technology would be used in real life (without prompts).

Analyzing the Results

The output of a usability test is a recording of the person navigating the screen and thinking aloud, along with your own notes of what you observed (if you ran a moderated test). As with other research methods, just one usability test is not super useful. You don't know if this person was an outlier and not really representative of a typical user. In order to analyze usability tests, you need to run multiple tests and compare the results. As with emotional journey mapping, you can look at your results in the aggregate and pay attention to common pain points or frustrations: Where did participants tend to get confused or frustrated? Did multiple users think aloud similarly, explaining why they were feeling confused or frustrated, at certain points of completing the task?

Usability test results in the UX industry are typically reported as recommendations for fixing specific problems with a digital product, so that the next iteration of the design can be more user-friendly and efficient. In academia, usability tests with different groups of people (based on identity, ability, etc.) can be used to empirically show existing power imbalances and biases in technology, particularly when thinking about who technology is designed for.

Case Study

Steve Krug is a usability consultant who has written numerous books on the topic of UX and specifically usability. In 2010, he wrote the book *Rocket Surgery Made Easy: The Do-It-Yourself Guide to Finding and Fixing Usability Problems* to show how usability testing is very simple. All it entails is observing someone using a digital product and talking through how they use it.[2]

Krug created a video of a demo usability test as a companion to his book, and you can watch it here! www.youtube.com/watch?v=1UCDUOB_aS8

In the video, Krug shows a screen recording of a participant completing various tasks on the Zipcar website (Zipcar is a car-sharing company where members pay hourly or daily rates to rent cars, along with their membership). In the video, Krug starts off by welcoming the participant and explaining the test. Next, the participant is prompted to provide a narrative about the home page, to see if it's clear to users what this site is for. The participant states that she knows this site is for Zipcar and that she knows what that is (renting cars), but that the bow on the picture of the car is confusing, as it makes her feel like she is *buying* a car. Krug could take this insight back to the content team and have them change the image on the front page to a car without a bow, to make it clearer that users will be renting a car and not buying one.

The participant then walks through the website, thinking out loud as she does so. She highlights certain phrases and exclaims "I don't know what this means." Once again, Krug could use this as an actionable insight to change the wording on the site to be clearer. Krug prompts the participant loosely throughout, asking her more questions about certain parts of the home page. Next, Krug takes the participant through specific tasks using scenarios prompting her to move through the site in particular ways (e.g., explore your options for using Zipcar for occasional trips) and observes her working through these tasks. At the end of the demo, Krug highlights three problems that the usability test made clear and then he suggests improvements to the site based on these insights. Watch the full demo, and make notes of what you think the three biggest usability problems are and how you would suggest fixing them.

Steps

1. Create (or have someone on your team, like a UX designer, create) a wireframe or prototype of what you want to test.
2. Write out two to three tasks that you would like the participants to complete on your website or app.
3. Decide on the type of test you will run:

   a. Moderated versus unmoderated

      If you choose to run unmoderated tests (without you being physically present), you need to make sure that the instructions and task lists are clearly written out for the participant, so they can understand what they need to do without needing clarification.

   b. Remote versus in-person

      i. For both types of tests, you should record the session. If you are in-person, you can sit next to the person and watch the screen over their shoulder as they work.
      ii. If you are remote, you can connect via Zoom (or another video communication software) and ask the participant to share their screen with you. If using Zoom, you will be able to see what is happening on the participant's screen, as well as being able to see their face, which is useful for analyzing nonverbal cues (such as a frown signifying confusion).

4. Start the test by welcoming the user and explaining what will happen during the test (remember, if you are conducting an unmoderated session, this all has to be written down for the participant to read on their own. Or you could record yourself introducing the test and providing the instructions and have the participant watch the video on their own). Make sure

to clarify that you are not testing *them*, but the website/app. There are no right or wrong answers! Explain that you will ask them to think aloud as they're moving through the tasks, telling you what they're doing and why they're doing it.

5. First, ask the participant about the home page. You might include questions such as: What is this page for? What could you do here? What stands out to you about this page? This tests first impression messaging resonance of your website or app.

6. Next, ask the participant to move through tasks on the site, such as finding something specific. Make sure to remind the participant to think out loud as they're completing the tasks (if you are running an unmoderated session, you can have popover reminders flash on the screen).

7. When the test is finished, thank participants for their time.

8. Analyze the results of multiple tests, paying attention to frequent pain points.

9. Report on the top usability problems your test showed and provide recommendations to the content creation and UI design teams to fix these problems.

10. Repeat with subsequent iterations, refining the design periodically.

Discussion Questions

1. Usability testing and contextual inquiry are similar in that both encompass talking with and observing users as they navigate through a digital space. How are they different? What different Research Questions might you answer using contextual inquiry than usability testing?

2. What are the pros and cons of in-person versus remote usability tests? What are the pros and cons of moderated versus unmoderated usability tests?

3. When do you think additional tools like eye movement tracking and heat maps are necessary? What are the pros and cons of including these in a usability study?

Notes

1. John Brooke, "SUS: A Quick Usability Scale," *Usability Evaluation in Industry* 189, no. 194 (1996).
2. Steve Krug, *Rocket Surgery Made Easy: The Do-It-Yourself Guide to Finding and Fixing Usability Problems* (Indianapolis: New Riders, 2009).

20 A/B Testing

A/B Testing is the most "quantitative" method included in this textbook. However, it is an extremely straightforward, simple, and quick way to know if your prototype should be implemented, and when talking about popular UX research methods, it is impossible to leave out. A/B testing is named as such because it is essentially a comparative test where you ask participants to choose "A" or "B," whether it's a specific feature (e.g., blue button versus red button) or specific language in content (e.g., using "students" versus "learners" in web copy). Usually consisting of one either/or question and one open-ended "why?" question, A/B testing is cheap and quick, but telling and useful for understanding users' preferences.

Quick Tips	
Tools	Google Forms, Survey Monkey
Use When	When you want to compare your new prototype to the current version or when you want to compare two new prototypes. Your research is about finding out the most effective way messaging for a particular target audience, for example, in health communication campaigns.
Design Thinking Stage	Test

The Method

Toward the end of the Design Thinking process, you will finally have some idea as to how you can resolve your research problem. You should now have a prototype that incorporates the small changes you have built through Empathize, Define, and Ideate methods. You have some ideas about different design or content options for the prototype, but how will you know which versions users prefer? Or, if you are building an improved version of a product that already exists, how do you know which version users prefer—your new one or the original? To answer these questions, you can use one of the most popular

DOI: 10.4324/9781003181750-23

UX methods: A/B testing. A/B testing places two options next to one another and asks participants to choose: "A" or "B"?

You can see an example of this in Figure 20.1. There are two options for a clothing site. Upon first glance they look quite similar, but you will notice that the first option's button says "BUY NOW," while the second option's button says "ORDER NOW."

There are two main ways to organize an A/B test. In the first scenario, "A" and "B" are both prototypes that you have created. Perhaps you found, through your research, that there are two slightly different ways a problem could be solved. In this case, you would create two new prototypes, each reflecting the small change. Then, you would ask participants to choose the prototype that they found to be better, as relevant to your study—more efficient, easier to use, more enjoyable, more aesthetically pleasing, etc.

In the second scenario, one example presented to the participants is your prototype and the other is how the product currently functions. For example, let's say you are working to redesign Gmail's icon. In this scenario, you would present participants with what the icon currently looks like and the new version you

Figure 20.1 An example of A/B test options.

have designed based on data from previous Design Thinking stages. As with the first scenario, you still ask participants which is "better," as well as why.

Calling it A/B testing can be confusing because it may lead you to believe that "A" must always be the same example and "B" must always represent the other example. However, changing the order of options for your participants is very important. You may have learned about order bias in a traditional research methods course. Order bias occurs when you present each participant with the options in the same exact order. Participants may be more likely to choose "A" because it is first and they view it as more important. Or maybe they choose "B" just because it is the last one they viewed, thus it is more present in their minds. Clearly, these can no longer be considered valid results because you haven't really tested the problem in question. So, even though this method is called A/B testing, the "A" and "B" shouldn't be the same for every participant. Instead, be sure to randomly switch which version is "A" and which is "B" each time a participant completes the study.

During A/B testing you can use either static images or working prototypes that allow users to actually interact with the interface. Usually a task handed off to UX Designers on your team, two or more slightly different working prototypes are created, using software like Figma or Adobe XD. Importantly, you will want the experience to be as close as possible to how the participant would actually use the product. For example, if the product being tested is a mobile app, the participants should complete the A/B test on their own smartphone.

A/B tests can be carried out in-person or virtually. In-person you can show a participant "A" and "B," learn which is their preference, and talk to them about why they chose that option. If the prototypes are static images, like the Gmail icon example, you can just show the participants the two images. To ensure order bias is not present, each participant should be shown the two options in different orders, randomly. This process could be set up ahead of time by using a random number generator that outputs, randomly, a "1" or "2" denoting which participant will see which option first. If it is a working prototype created using a program like Figma, it is best to provide the links to the two options, still randomly labeling them "A" and "B." The next best step is to then have your participants view the two links on their own device. If not, they can use a phone, tablet, laptop, or desktop computer that you may have in the lab or have brought with you to the testing site.

To conduct virtual A/B tests, you can also use online survey tools like Google Forms and SurveyMonkey. Setting up an A/B test survey is straightforward. You need only include three questions. First, you will want some identifying information. Perhaps it is an email address or a full name. This way you can link each participant back to their demographic information. (If you are only working with these participants for this one A/B test, then you will want to collect any relevant demographic information within the survey as well, so you can use this information in the creation of personas or other future design activities.) Second, you will include the two A/B options as images or links (more on this below). Some survey forms provide a checkbox that switches on a tool that randomly shows one of the options first. (If your survey service does not do this, you can create two nearly identical surveys that only differ in their order of the two options. Then you can randomly select which link goes out to each participant. Again, this is to prevent order bias.) Third, you will want to include some type of open-ended, "why?" question in your survey. You may just want to ask "Why did you choose this option?" Or, you may want to ask a more specific "why" question.

If you are using static images as your two options, you can simply upload those images to the survey. Then, participants can view them within the site and choose. If you are testing working prototypes, then you will most likely include two links, one to each of the options, in your survey. Participants can then click on each, one at a time, try out the two options, and then go back to your survey to select "A" or "B" and to explain why. Remember, you should ask participants to use the device for which you are targeting the prototype. If you are testing a mobile interface, but participants are using their desktop computers, you may not get as authentic results as if all participants used their mobile phones to check out your prototypes.

Use When

A/B testing is best used when you are ready to test a new tool or functionality that you have developed after collecting data from participants in the Empathize stage. A/B testing is great for comparing two prototypes or your new prototype to an already existing product. Be careful to ensure that your "A" and "B" versions are only testing one or two small changes at a time. If there is too much variability, it will be difficult to determine why participants chose one option over the other. So, if you change the color of a button, for example, make sure the wording, shape, placement, etc. stays the same.

In academic research, A/B testing can be used in a similar manner—to better understand which option your participants prefer. For instance, if you study any type of persuasive communication, such as health communication or political communication, you can use A/B testing to understand what type of wording or content works best in campaigns. A/B testing can be used in tandem with traditional methods like qualitative surveys and interviews to better understand attitudes or opinions towards different choices on the topic under study. You can even present an A/B choice during a focus group and see how the participants discuss which is preferred.

What You Need

To begin, you will need two options that you are testing against each other. Again, these two options may be two new versions of your app or website (or more likely a small part of it), or one may be an existing product while the other is your newly developed version. These can be static images if the change is aesthetic, or they may be working prototypes that allow participants to click, drag, type, and swipe as if they are really using the app or website. In most cases, prototypes are delivered to you by a UX designer, but in some cases, UX researchers also design prototypes using Design programs like Figma and Adobe XD.

Next, you will need a way of getting options "A" and "B" to your participants. If working in person, you may just be able to show your participants the two static options. Or, if testing working prototypes, you can provide them with links to each. While it is best to have participants use their own devices so the experience is as close to their everyday interactions with your product, you may have to provide them with the device you wish them to view "A" and "B" on.

If testing virtually, it is best to use a survey service like Google Forms or SurveyMonkey. Create a survey that collects whatever demographic information you may need, asks participants to choose "A" or "B" (either with static images right in the survey or with links to your working prototypes), and includes some kind of "why" question to better understand their selection.

Analyzing the Results

In most cases, analyzing A/B test results is fairly straightforward. You will first take note of how many participants chose "A" and how many chose "B." You will need to be sure that you normalize your data since about half of your participants viewed option one as "A" and option two as "B." In other words, be sure you don't count all "As" as the same! Once you have your count as to the preferred option, you can begin to thematically analyze the responses to your "why" question.

If your A/B test is comparing a new prototype to the existing version of a product, the analysis is even easier. You generally now have data-driven support for making that change to your app or website because your participants have chosen that as a better option than the current version. However, it is still important to analyze the "why" responses. In some cases, participants may have selected your new prototype simply because it is different from what they are used to, especially if they are already frustrated with current designs or processes. The opposite may also be true: participants may be hesitant to choose your prototype over the current version simply because they are accustomed to the current one and don't want to go through any perceived learning curve to continue to use your product. Again, this is why it is important to read through the "why" responses.

If your A/B test is comparing two new prototypes, the analysis portion is a bit more involved. In some cases, there may be an overwhelming majority that has chosen one option over another. Still read over your "why" responses, but it is likely safe to say that users will generally enjoy the update. In other cases, however, selections may be closer to a 50/50 split. Here is where your thematic analysis skills really come in handy. Why have participants chosen certain options? Should you now return to another Design Thinking stage and collect more data?

Finally, reading through "why" responses is also more beneficial for researching your product more generally. You can learn more about how users (or potential users) feel about specifics of your app or website. These insights can help with later analysis and even with preparing future studies,

Case Study

At the end of 2007, Dan Siroker, now the founder and CEO of Scribe AI, worked as a Google product manager. But, he took a leave of absence after meeting Barack Obama, then a presidential candidate, at a talk at Google head-quarters. As an Obama campaign digital advisor, Siroker introduced Obama to A/B testing.[1]

Obama's digital team was having a big issue with turning site visitors into subscribers—getting a visitor to subscribe meant the campaign team collected an email, and once emails were collected subscribers could turn into donors. The original website began with a splash page—the introductory page that pops up whenever any page of your website is visited—that included a turquoise-shaded photo of Obama and a bright-red "sign up" button. Although the page got many visitors, not many clicked the "sign up" button.[2]

Working with the team, Siroker pulled out the two main parts of the splash page—the image and the button. Using A/B testing, the team tested different button wording including "Learn More," "Join Us Now," and "Sign Up Now." They also tested a few different visuals—a black-and-white photo of the Obama family and a video of Obama speaking at a rally. The "Learn More" button led to 18.6% more signups. The black-and-white photo outperformed all the other visuals by 13.1%. Interestingly, the team's "instinct" was that the video would be much more effective. But, it was actually 30.3% worse! This is a perfect example of how researchers are *not* the users. Changes should be data-driven. Together, using the new black-and-white family photo and the "Learn More" button, the team reached 40% more signups. It was estimated that about $75 million of campaign money was brought in due to the site's UX changes.[3]

Steps

1. Receive your prototype(s) from the UX designer, or create them yourself using a tool like Figma or Adobe XD.
2. Decide if you will conduct the A/B tests in-person or virtually.

   a. If conducting in-person, prepare the materials. These include the pro-totypes you are testing and possibly a device for your participant to view them on if they are not using their own. But, remember, it is best to have participants use their own devices so that the test is as close as possible to their day-to-day usage.

b. If conducting virtually, create your survey using a resource like Google Forms or SurveyMonkey. Remember that if you are using static images you can include them right in the survey. If you are using working prototypes, you should link the participants to each in the survey.

3. To prevent order bias, be sure to randomize the order that the participants see the options.

4. If working in-person, be sure to explain to your participants the process they should follow before they begin. If working virtually, don't forget to include any relevant instructions at the top of the survey. It is usually a good idea to put the instructions on their own, introductory page. Participants can then click a "next" button to move on to the actual test.

5. Once all of your participants have completed the test, analyze the results. Remember to look at the quantitative aspects (how many picked which option) as well as perform a thematic analysis of your qualitative, "why" responses.

Discussion Questions

1. In what situations would testing static images be better than testing working (interactive) prototypes? In what situations would testing working prototypes be better than testing static images?

2. Although A/B testing is considered quite popular in the UX community, some think it is overused or used improperly. Can you think of a scenario for which A/B testing would not be appropriate? Suggest another test method discussed in this book and explain why it would be better than an A/B test in your example.

3. A/B tests are rarely used in academic settings. Why do you think this is the case? Can you think of a Research Question for which an A/B test could be combined with interviewing, focus groups, or surveys?

Notes

1. Brian Christian, "The A/B Test: Inside the Technology That's Changing the Rules of Business," *Wired*, April 25, 2012, www.wired.com/2012/04/ff-abtesting/.
2. Ibid.
3. Ibid.

Bibliography

Apple. "Human Interface Guidelines." Accessed July 18, 2021. https://developer.apple.com/design/human-interface-guidelines/.

Baker, Monya. "1,500 Scientists Lift the Lid on Reproducibility." *Nature News* 533, no. 7604 (2016). www.nature.com/articles/533452a.

Bales, Elizabeth, Timothy Sohn, and Vidya Setlur. "Planning, Apps, and the High-End Smartphone: Exploring the Landscape of Modern Cross-Device Reaccess." In *International Conference on Pervasive Computing*, edited by Kent Lyons, Jeffrey Hightower, and Elaine M. Huang, 1–18. Berlin: Springer-Verlag, 2011.

Bergin, Thomas J. "The Origins of Word Processing Software for Personal Computers." *PC Software: Word Processing for Everyone* (2006): 32.

Bivens, Rena, and Anna Shah Hoque. "Programming Sex, Gender, and Sexuality: Infrastructural Failures in the 'Feminist' Dating App Bumble." *Canadian Journal of Communication* 43, no. 3 (2018): 441–59. doi.org/10.22230/cjc.2019v44n3a3375.

Brooke, John. "SUS: A Quick Usability Scale." *Usability Evaluation in Industry* 189, no. 194 (1996): 4–7.

Bunderson, Eileen D., and Mary Elizabeth Christensen. "An Analysis of Retention Problems for Female Students in University Computer Science Programs." *Journal of Research on Computing in Education* 28, no. 1 (1995): 1–18.

Burns, Matt. "BMW's Magical Gesture Control Finally Makes Sense as Touchscreens Take Over Cars." November 4, 2019. https://techcrunch.com/2019/11/04/bmws-magical-gesture-control-finally-makes-sense-as-touchscreens-take-over-cars/?guccounter=1.

Butler, Judith. *Gender Trouble*. New York: Routledge, 1999.

CAST. "The UDL Guidelines." 2018. https://udlguidelines.cast.org/.

Christian, Brian. "The A/B Test: Inside the Technology That's Changing the Rules of Business." *Wired*, April 25, 2012. www.wired.com/2012/04/ff-abtesting/.

Cooper, Alan. "The Long Road to Inventing Design Personas." *OneZero*, February 4, 2020. https://onezero.medium.com/in-1983-i-created-secret-weapons-for-interactive-design-d154eb8cfd58.

Dam, Rikke Friis, and Teo Yu Siang. "Personas: A Simple Introduction." *Interaction Design Foundation*, January 2021. www.interaction-design.org/literature/article/personas-why-and-how-you-should-use-them.

David, Gaby, and Carolina Cambre. "Screened Intimacies: Tinder and the Swipe Logic." *Social Media + Society* 2, no. 2 (2016). https://doi.org/10.1177/2056305116641976.

Dennie, Danielle, and Susie Breier. "The Wind Beneath My Wings: Falling in and out of Love with an Online Library Research Skills Tutorial." *Concordia Library*, 2019.

https://spectrum.library.concordia.ca/985363/1/breier-dennie-forum-2019-final-20190426.pdf.

DeVito, Michael A., Ashley Marie Walker, and Jeremy Birnholtz. "'Too Gay for Facebook': Presenting LGBTQ+ Identity Throughout the Personal Social Media Ecosystem." *Proceedings of the ACM on Human-Computer Interaction* 2, no. CSCW (2018): 1–23.

Dosono, Bryan, Jordan Hayes, and Yang Wang. "'I'm Stuck!': A Contextual Inquiry of People with Visual Impairments in Authentication." In *Eleventh Symposium on Usable Privacy and Security ({SOUPS} 2015)*, 151–68. 2015. https://www.usenix.org/conference/soups2015/proceedings/presentation/dosono

Doubleday, Alison. "Use of Card Sorting for Online Course Site Organization Within an Integrated Science Curriculum." *Journal of Usability Studies* 8, no. 2 (2013): 41–54.

Eden, Colin. "Cognitive Mapping." *European Journal of Operational Research* 36, no. 1 (1988): 1–13.

Elsbernd, Abbey, Maiken Hjerming, Camilla Visler, Lisa Lyngsie Hjalgrim, Carsten Utoft Niemann, Kirsten Arntz Boisen, Jens Jakobsen, and Helle Pappot. "Using Cocreation in the Process of Designing a Smartphone App for Adolescents and Young Adults With Cancer: Prototype Development Study." *JMIR Formative Research* 2, no. 2 (2018). http://formative.jmir.org/2018/2/e23/.

English, William K., Douglas C. Engelbart, and Melvyn L. Berman. "Display-Selection Techniques for Text Manipulation." *IEEE Transactions on Human Factors in Electronics* 1 (1967): 5–15.

ENIAC. "ENIAC Programmers Project." 2021. http://eniacprogrammers.org/.

Gagne, Yasmin. "Nike's New Concept Store Feeds Its Neighbors' Hypebeast and Dad-Show Dreams." *Fast Company*, July 12, 2018. www.fastcompany.com/90201272/nikes-new-concept-store-feeds-its-neighbors-hypebeast-and-dad-shoe-dreams.

Google. "Guidelines." Accessed July 18, 2021. https://material.io/design/guidelines-overview.

Harding, Sandra G. *The Science Question in Feminism*. Ithaca: Cornell University Press, 1986.

Hasso Plattner Institute of Design at Stanford. *An Introduction to Design Thinking: Process Guide* (2013). https://web.stanford.edu/~mshanks/MichaelShanks/files/509554.pdf.

Heslin, Peter A. "Better Than Brainstorming? Potential Contextual Boundary Conditions to Brainwriting for Idea Generation in Organizations." *Journal of Occupational and Organization Psychology* 82, no. 1 (2009): 129–45.

Hicks, Mar. "When Did the Fire Start?" In *Your Computer Is on Fire*, edited by Thomas S. Mullaney, Benjamin Peters, Mar Hicks, and Kavita Philip, 11–25. Cambridge: MIT Press, 2021.

Holtzblatt, Karen, and Hugh Beyer. *Contextual Design: Defining Customer-Centered Systems*. Amsterdam: Elsevier, 1997.

Hovland, Ingie. *Successful Communication: A Toolkit for Researchers and Civil Society Organisations*. London: Overseas Development Institute, 2005. https://citeseerx.ist.psu.edu/viewdoc/download?doi=10.1.1.123.1057&rep=rep1&type=pdf.

IDEO Design Thinking. "Design Thinking Defined." Accessed July 18, 2021. https://designthinking.ideo.com/.

IDEO Design Thinking. "History." Accessed July 18, 2021. https://designthinking.ideo.com/history.

Interaction Design Foundation. "Design Thinking." Accessed July 18, 2021. www. interaction-design.org/literature/topics/design-thinking.

Interaction Design Foundation. "What Is User Interface Design?" Accessed July 18, 2021. www.interaction-design.org/literature/topics/ui-design.

International Service Design Institute. "The Story of the Journey Map—The Most Used Service Design Technique." 2020. https://internationalservicedesigninstitute.com/the-story-of-the-journey-map-the-most-used-service-esign/.

IoT Agenda. "Internet of Things (IoT)." Accessed July 18, 2021. https://internetofthing-sagenda.techtarget.com/definition/Internet-of-Things-IoT.

ISO. "ISO 9241–11:2018(en)." 2018. www.iso.org/obp/ui/#iso:std:iso:9241:-11:ed-2:v1:en.

Jansen, Bernard J. "The Graphical User Interface." *ACM SIGCHI Bulletin* 30, no. 2 (1998): 22–26.

Jewell, Lisa. "User Research—What's Tomato Ketchup Got to Do With It?" *UX Planet*, May 14, 2018. https://uxplanet.org/user-research-whats-tomato-ketchup-got-to-do-with-it-758bfb536ca3.

Kapor, Mitchell. "A Software Design Manifesto." In *Bringing Design to Software*, edited by Terry Winograd, 1–6. New York City: ACM Press, 1996.

Krug, Steve. *Rocket Surgery Made Easy: The Do-It-Yourself Guide to Findings and Fixing Usability Problems*. Indianapolis: New Riders, 2009.

Light, Ben, Jean Burgess, and Stefanie Duguay. "The Walkthrough Method: An Approach to the Study of Apps." *New Media & Society* 20, no. 3 (2018): 881–900. https://doi.org/10.1177/1461444816675438.

Linzmayer, Owen W. *Apple Confidential 2.0: The Definitive History of the World's Most Colorful Company*. San Francisco: No Starch Press, 2004.

Lubar, Steven. " 'Do Not Fold, Spindle or Mutilate': A Cultural History of the Punch Card." *Journal of American Culture* 15 (1992): 43–43.

McLeod, Caitlin, and Victoria McArthur. "The Construction of Gender in Dating Apps: An Interface Analysis of Tinder and Bumble." *Feminist Media Studies* 19, no. 6 (2019): 822–40. https://doi.org/10.1080/14680777.2018.1494618.

Montello, Daniel R. "Cognitive Map-Design Research in the Twentieth Century: Theoretical and Empirical Approaches." *Cartography and Geographic Information Science* 29, no. 3 (2002): 283–304.

Morville, Peter. "User Experience Design." *Semantic Studios*, June 21, 2004. https://semanticstudios.com/user_experience_design/.

Myers, Brad A. "A Brief History of Human-Computer Interaction Technology." *Interactions* 5, no. 2 (1998): 44–54.

NASA. "Katherine G. Johnson." May 25, 2017. www.nasa.gov/feature/katherine-g-johnson.

National Museum of American History. "Xerox 8010 Star Information System." Accessed July 18, 2021. https://archive.org/details/byte-magazine-1981-09/page/n59/mode/2up.

Nielsen, Jakob. "10 Usability Heuristics for User Interface Design." *Nielsen Norman Group*, November 15, 2020. www.nngroup.com/articles/ten-usability-heuristics/#poster.

———. "A 100-Year View of User Experience." *Nielsen Norman Group*, December 24, 2017. www.nngroup.com/articles/100-years-ux/.

Nielsen, Jakob, and Rolf Molich. "Heuristic Evaluation of User Interfaces." *Proceedings of the SIGCHI Conference on Human Factors in Computing Systems* (1990): 249–56. https://doi.org/10.1145/97243.97281.

Noble, Safiya Umoja. *Algorithms of Oppression*. New York: NYU Press, 2018.

Norman, Don. "Emotion & Design: Attractive Things Work Better." *Interactions* 9, no. 4 (2002): 36–42.

Norman, Don, and Jakob Nielsen. "The Definition of User Experience (UX)." *Nielsen Norman Group*. Accessed July 18, 2021. www.nngroup.com/articles/definition-user-experience/.

Norman, Donald A. "The Way I See It: Systems Thinking: A Product Is More Than the Product." *Interactions* 16, no. 5 (2009): 52–54.

O'Neil, Cathy. *Weapons of Math Destruction*. New York: Crown Publishing Group, 2016.

PCMag. "Interface." Accessed July 18, 2021. www.pcmag.com/encyclopedia/term/interface.

Ritter, Marli. "Heuristic Analysis of Yuppiechef.com: A UX Case Study." *UX Collection*, October 7, 2018. https://uxdesign.cc/ux-case-study-heuristic-analysis-of-yuppiechef-com-c92098052ce4.

Roberts, Steven K. "The Xerox Alto Computer." *BYTE Magazine*, September 1981. https://archive.org/details/byte-magazine-1981-09/page/n59/mode/2up.

Rossi, Cesare, Flavio Russo, and Ferruccio Russo. *Ancient Engineers & Inventions*. Dordrecht: Springer, 2009.

Salazar, Kim. "Contextual Inquiry: Inspire Design by Observing and Interviewing Users in Their Context." *Nielsen Norman Group*, December 6, 2020. www.nngroup.com/articles/contextual-inquiry/.

Schooler, Jonathan W. "Metascience Could Rescue the 'Replication Crisis'." *Nature News* 515, no. 7525 (2014). https://doi.org/10.1038/515009a.

Section508.gov. "Accessibility Testing for Websites and Software." Accessed July 18, 2021. www.section508.gov/test/web-software.

Shaw, Adrienne. "Encoding and Decoding Affordances: Stuart Hall and Interactive Media Technologies." *Media, Culture, & Society* 39, no. 4 (2017): 592–602.

Siltanen, Rob. "The Real Story Behind Apple's 'Think Different' Campaign." *Forbes*, December 14, 2011. www.forbes.com/sites/onmarketing/2011/12/14/the-real-story-behind-apples-think-different-campaign/?sh=41ac8a6362ab.

Spencer, Donna, and Todd Warfel. "Card Sorting: A Definitive Guide." *Boxes and Arrows*, April 7, 2004. https://boxesandarrows.com/card-sorting-a-definitive-guide/.

Stephenson, William. *The Study of Behavior: Q-Technique and Its Methodology*. Chicago: University of Chicago Press, 1953.

Tolman, Edward C. "Cognitive Maps in Rats and Men." *Psychological Review* 55, no. 4 (1948): 189–208. https://doi.org/10.1037/h0061626.

Torres de Souza, Mady, Olga Hording, and Sohit Karol. "The Story of Spotify Personas." *Spotify Design*, March 2019. https://spotify.design/article/the-story-of-spotify-personas.

U.S. Access Board. "About the ICT Accessibility 508 Standards and 255 Guidelines." Accessed July 18, 2021. www.access-board.gov/ict/.

Usability.gov. "Scenarios." Accessed July 18, 2021. www.usability.gov/how-to-and-tools/methods/scenarios.html.

VanGundy, Arthur B. "Brain Writing for New Product Ideas: An Alternative to Brainstorming." *Journal of Consumer Marketing* 1, no. 2 (1984): 67–74. https://doi.org/10.1108/eb008097.

Vimeo. "Smart Design: The Breakup Letter." 2010. https://vimeo.com/11854531.

Wachter-Boettcher. *Technically Wrong*. New York: WW Norton & Company, 2017.

Wajcman, Judy. "Feminist Theories of Technology." *Cambridge Journal of Economics* 34, no. 1 (2010): 143–52. https://doi.org/10.1093/cje/ben057.

Winner, Langdon. "Do Artifacts Have Politics?" *Daesalus* 109, no. 1 (1980): 121–36. www.jstor.org/stable/20024652.

Zemke, Ron, and Chip R. Bell. *Service Wisdom: Creating and Maintaining the Customer Service Edge*. Minneapolis: Lakewood Books, 1989.

Appendix A

Sample Research Plan Template

Title of Research Project
Researcher/Author
Date

Background

Relevance: Why are you doing this project?

- Why is it important/timely?
- What problem are you trying to solve?

Prior Research: What research has been done on the topic? Summarize prior findings (this is the Literature Review in academic research)

Research Questions or Research Objectives

What question are you answering/What are the objectives for this study?

Participants

How many people will you talk to?

Where/how will you find them (recruitment)? e.g., through social media posts

Who are they? (demographics—age, gender, location, marital status, income, etc.)

Criteria for participation in this project—what types of people are you trying to understand? (e.g., must haves: familiarity with a technology, be a pet owner, drink coffee)

Method

What specific method will you use? (e.g., contextual inquiry, usability testing)

Where will the study take place?

How will you answer your question or fulfill your objectives?

- Include all study details, such as the physical setup and questions you will ask the participants (interview guide)

Timeline

Details on timeline for:

- Participant recruitment
- Data collection
- Data analysis
- Presentation of findings

References/Appendixes

What other documents are useful to understanding this research plan? For example, recruitment ads (this is a bibliography in academic research)

Appendix B

Academic Report Example

Normative Interfaces: Affordances, Gender, and Race in Facebook
Angela M. Cirucci, PhD

This article was first published by SAGE in *Social Media + Society*:
Angela M. Cirucci, *Normative Interfaces: Affordances, Gender, and Race in Facebook*, Social Media + Society (vol. 3, issue 2) pp. 1–10. Copyright © 2017 (Angela M. Cirucci). doi:10.1177/2056305117717905.

Normative Interfaces

AFFORDANCES, GENDER, AND RACE IN FACEBOOK

With each new technological advancement comes a declaration of some "great social equalizer" (boyd 2014). When the internet was first entering households, a common belief was that its integration would bring a cultural and social shift. These sentiments were guided by the idea that virtual communities allowed users to leave their bodies behind; users met new people and experimented with their identities (Rheingold 1996; Turkle 1995). Prejudices were assumed to soon be a thing of the past—race, gender, and physical appearances would no longer be delineating factors.

Today, these utopian visions are criticized for their optimism. It seems that technologies cannot solve social issues and perhaps even work to emphasize social divisions (boyd 2014). The prejudices that we learn offline are likely to journey with us into digital spaces. Although Facebook, for example, allows users to connect to people in new ways, it also reinforces existing networks and norms transferred from offline spaces. In other words, Facebook relationships are "anchored" and compel users to value nonymous (Zhao, Grasmuck, and Martin 2008), and perhaps even anti-anonymous (Cirucci 2015), identifications over anonymous ones.

The goal of this study is to better comprehend, as they relate to race and gender, social network site affordances and related interpretations of gender and race identities. Employing Facebook as a case study, I seek to develop answers to questions such as: How familiar are users with Facebook's tools and functionalities? How are issues of gender and race represented through Facebook? How do users conceive of gender and race?

Affordances

The use of "affordance" as a noun was coined by Gibson in his 1979 book *The Ecological Approach to Visual Perception.* Writing about animals and their biological environments, Gibson uses his theory of affordances to explain that, to really understand where and how animals live, we must comprehend how animals visually perceive of what their environments offer them:

> I mean it by something that refers to both the environment and the animal in a way no existing term does. It implies the complementarity of the animal and the environment. . . . Affordance cuts across the dichotomy of subjective-objective and helps us to understand its inadequacy. It is equally a fact of the environment and a fact of behavior. It is both physical and psychical, yet neither. An affordance points both ways, to the environment and to the observer.
>
> (pp. 119, 121)

In other words, there is a symbiotic relationship between the animal and its environment.

Gibson continues by explaining what has happened now that humans have added on to the environment—the shapes and substances of our world have been changed because humans want to make more available what benefits them. The large, and important, change to the conception of affordances then is that human-made objects are no longer neutral. The symbiotic process of animal-environment takes on a third party—designers.

Gibson's original argument is actually situated in *ignoring* the material makeup of objects and instead analyzing *affordances*—what the object can do, is supposed to do, and is promoted as doing. At the time of his writing, Gibson believed that people paid too much attention to the dimensions or physical qualities of things—their smaller pieces, their material makeup. Instead, he wanted researchers to focus on what objects may *afford* animals in the symbiotic relationship of animal-environment. We have perhaps come full circle, often omitting material analyses of mediated structures. And, media technologies present different issues than trees or flowers because they are made by humans and are thus not neutral. Indeed, the general social network site user is rarely, if ever, provoked to think about the material makeup of the site.

In sum, to apply Gibson's theory of affordances is to focus on the negotiation that exists at the intersection of user, interface, *and* designer; affordances do not exist without interaction (Nagy and Neff 2015). Without considering all three pieces, we miss the dynamic relationship that all users have with mediated technologies. Users display a stark differentiation between their practices and the actual interfaces. Theoretically, we can think of things like features, apps, and devices as distinct objects, but users make sense of technological systems in a variety of ways that span and conflate these objects and interfaces (McVeigh-Schultz and Baym 2015).

McVeigh-Schultz and Baym (2015, 2) define affordances as the "perception between bodies and artifact." Similarly, Van Dijck (2013) argues that platforms are techno-social constructs that are some aggregation of technology, content, and users. However, neither explicitly states the new third prong—designers. Thus, I argue, to study affordances is to study the negotiation that happens between users *and* designers, through the created interface. While it is important to view interfaces and digital objects more like mediators instead of intermediaries (because they do much more than just move information but transform it) (e.g., Galloway 2013; Latour 2005), it is integral to also include investigations into how and why specific designer choices have implications for users.

As such, using affordance to mean "choice" or "constraint" is not helpful when examining the three-pronged negotiation (McVeigh-Schultz and Baym 2015). Instead, affordances are relational negotiations (e.g., Hutchby 2001a) that are constantly occurring as the cycle of designers designing and users using continues. While objects may visually convey action-capacities (Norman 1988), digital capacities can be hidden by designers in an effort to promote desired perceptions and uses (McVeigh-Schultz and Baym 2015).

In addition, because designers have the most power in affordance negotiation, their decisions work to prime, resist, and shape the ways users make sense of the technology (McVeigh-Schultz and Baym 2015). "When a programmer decides which gesture to render, then [they are] deciding not what to communicate, but what possible messages to allow; such decisions dictate the communication potential of a space" (Kolko 1999, 180). Interfaces are both semiotic and institutional structures (Giddens 1984) that influence how narratives are shared and shaped (Duguay 2016). As media technologies become more advanced, designers work harder to design "user-friendly" spaces that obscure how spaces actually function. Less space is provided to tinker and, as a general culture, we are encouraged to feel a passivity to technology (Gillespie 2006).

Previous research in online gaming has highlighted the importance of avatar design because of its strong tendencies to affect offline interactions (e.g., Kolko 1999). While this work was situated in games, social network site profiles are very similar to personalized avatars (e.g., Cirucci 2013). At sign-up, as with initial avatar creation before a game can be played, Facebook begins shaping user performances and experiences (Light and McGrath 2010). Facebook's "real name" policy, for example, along with a myriad of other design choices are generally in contention with identities that are fluid and complex (Lingel and Golub 2015).

Instead, Facebook is concerned with investing time and money into singular profiles (Lingel and Golub 2015). Some identity choices are inescapable while others are simply omitted (Zhao, Grasmuck, and Martin 2008). These flat profiles produce data that conflate identity performances and contexts and feed Facebook data that can then be repurposed as both targeted ads for users and valuable fodder for third-parties ready to cross-reference databases and build even more dynamic "profiles" of people. However, these databases are never perfect and privilege one point of view while muzzling many others (Light 2007). Facebook has thus become a polemical and political site of analysis (e.g., Bowker and Star 1999).

With the above in mind, this study aims to better understand the non-neutral interfaces that guide user identity perception on Facebook. Specifically, I explore identifications related to gender and race in an effort to better realize how and why Facebook potentially proliferates social division and misrepresentation. Because digital technologies are created by people, they are necessarily couched in political, economic, and cultural powers (e.g., Winner 1980). Therefore, a look into designer choices, paired with users' perceptions and daily usage habits, offers insight into Facebook's affordances.

Methods

Nagy and Neff (2015) outline a new way of conceptualizing affordances—imagined affordances. The notion of imagined affordance takes into account the experiences not consciously realized by users. They argue that users realize specific affordances, not some full set as presented by each space. On the other hand, McVeigh-Schultz and Baym (2015) found that people make sense of material structures at different, nested layers and that this sense-making process does not involve speaking about material parts separately. Clearly, there is a disconnect between how we can better investigate structures, their design, and their implications for users.

In response to these two studies, this study was conducted in two parts: a structural discourse analysis and focus groups. Pairing an analysis of presented, non-neutral tools with users' experiences with these tools provides a dynamic look into the negotiation and interaction that is affordances.

Structural Discourse Analysis

As Galloway (2013) proposes, I viewed Facebook's interface not as a medium, but as a mediator that does not transport, but transforms, information (Latour 2005). Before examining identity performances through the interface, I first explored the ways in which Facebook mediates identifications. In January 2014,[1] I reviewed and catalogued all of Facebook's buttons, tools, and functionalities. I then conducted a discourse analysis, noting what the tools made possible, what was impossible, and what actions and perceptions were privileged over others. Further, mostly through gaining access to *The Zuckerberg Files*,[2] I searched news archives to include, when possible, when, why, and how interface changes were made.

In sum, I conducted a discourse analysis not of user-generated content but of Facebook's structure—a structural discourse analysis. My technique was similar to Duguay's (2015) walkthrough method wherein she notes parts of technologies' architectures such as content navigation tools, features, and buttons. However, the moral values embedded in Facebook (Light and McGrath 2010) were more deeply considered through the discourse analysis. My overall process was guided by Fairclough (1995) and partially inspired by Papacharissi's (2009) study of social network "geographies." I attempted to examine the social powers at play and how those in power attempt to control content and their structures. I view Facebook as a sociocultural system whose presented language and structure play a role in shaping identities, social relations, and systems of knowledge (Fairclough 1995, 55).

Focus Groups

After many readings of each tool and functionality, I conducted nine focus groups with college students (n=45) at a large, urban, east-coast university in the US. Beyond an analysis of Facebook's interface, I was interested in learning how everyday users identify and make sense of the site. Instead of just presenting architectural pieces as choices or constraints, I provided a space for informants to share their sense-making processes (McVeigh-Schultz and Baym 2015). As Duguay (2015) notes, employing only a walkthrough method is limiting because there is no inclusion of user perception. At the same time, only completing an analysis of user performances fails to recognize and consider the mandatory setting of the mediating interface.

Participants were 18–30 and declared their racial affiliations as: white (71%), Black (13%), Asian (9%), Latinx (4%), and Other (2%). Each focus group lasted between 45 minutes and 1.5 hours. Topics for conversation were derived from my structural discourse analysis findings. Informants were asked to share stories regarding how they have used different Facebook tools and were asked to speak about the extent to which they are aware of their non-neutrality. While some general questions were inspired by the discourse analysis, the focus groups were generally open-ended.

It is important to note that all participants were given a short demographics survey to complete. Spaces to define gender and race affiliations were open-ended. However, no participants included a gender affiliation beyond female or male. Focus groups were also mixed gender and race, which may have silenced those who may have felt their comments would have been seen as marginalized or "different." However, I chose focus groups because they are social and thus relevant when the content up for discussion is also social (Frey and Fontana 1993). As discussed in detail below, the makeup of my sample became a finding in and of itself, displaying how those in more privileged socio-cultural positions (or those who at least feel the need to speak in a way that conforms to the majority) negotiate affordances in very specific ways.

Findings

The following sections outline three main themes that emerged through the structural discourse analysis and focus groups. These three themes are rooted in Facebook's gender and race affordances. Each section outlines related functionalities, presents a brief discourse analysis, and shares participant experiences, as a way of exploring Facebook's affordances.

Digital Gender

In early 2014,[3] beyond binary options, Facebook provided US users 50+ gender affiliations.[4] It is safe to say that gender selection is important to the site; while on the About Page users can choose something other than a binary affiliation, at sign up new users still must choose from only "female" and "male." For Facebook, third party marketers, and database companies gender is a crucial demographic.

The study of digital gender is certainly not new. The importance of ascribing binary gender to virtual bodies has been integral to online worlds since early multi-user dungeons (MUDs), gaming spaces, and chat rooms. While some may have thought that users logged in and became "disembodied," it was quickly made apparent that electronic worlds are not separable from the physical self (e.g., Kolko 1999). In a study of Gaydar, a dating site for gay men, Light (2007) found that suggestions for users as they created their dating profiles were stereotypically masculine, omitting groups from drop-down boxes, specifically effeminate men. Through its digital structure, Gaydar users were pressured to conform to certain cultures with little room for resistance. This 2007 study is but one example of the ways in which designer choices of digital spaces make negotiation of gender affordances unequal and compel users to adhere to heteronormative expectations.

Although Facebook's About Page gender change occurred shortly before I spoke with my informants, only half were familiar with the additions, and only one had changed the gender option (from female to cisfemale). For my informants, and perhaps in line with Facebook's goals, the gender prompt is not perceived as a space for expression, but as a box to check that mirrors their birth certificates or medical forms.

> *Cheryl, an 18-year-old white female:*[5] Facebook's going to keep upgrading to try and make people want to go on it and feel more comfortable, like, they're accepted there. But, I just think that's ridiculous. Like, on your birth certificate I really doubt it's ever going to be more than just male or female.
>
> *Ryan A., a 19-year-old white male*: I think as soon as I got a Facebook I put it [gender selection], and my gender hasn't changed; though, I haven't really changed.

This brief excerpt from one of Ryan's stories nicely summarizes most of my informants' views. For my sample, changing gender affiliation on the site has nothing to do with avoiding stereotypes or enacting agency. Instead it is purely about filling in what is usually required on legal documents. For Facebook, this is defined as being "authentic" (putting in "correct" or "real" information). In reality, profile information that matches other identity performances, and thus buying habits, makes users more valuable.

When first introduced, Facebook users were not obligated to select their gender. It was an option, but never choosing one or the other was also possible, resulting in pronoun allotments becoming "they" and "their." Eventually, users who had chosen this neutral option received a message:

> Right now your mini-feed may be confusing. Please choose how we should refer to you. [user-first-name] edited *her* profile. [user-first-name] edited *his* profile.
>
> (McNicol 2013, 204)

On June 27, 2008, Naomi Gleit, Facebook Director of Product for Growth and Engagement, posted a Facebook note explaining the change. She claimed that translators were having some trouble.

> However, we've gotten feedback from translators and users in other countries that translations wind up being too confusing when people have not specified a sex on their profiles. People who haven't selected what sex they are frequently get defaulted to the wrong sex entirely in Mini-Feed stores.
>
> (Gleit 2008)

In her post, Gleit labels the affiliations "sex." This is how Facebook previously labeled the category. Gleit includes this in her post to connote a more objective sense of the identity label. Besides conforming to the English (US) delineation on the site, she is also implying that all users need to check off a sex, just as the hospital did when they were born. Later in the post, however, Gleit refers to the selection as "gender."

Some of my participants even noted that using the space to perform more than gender ascribed at birth is identification "overload."

> *Davina, a 19-year-old Black female:* If you are non-binary, then . . . whoever knows you would know that. Like, does that have to be the first thing that somebody sees on your profile?

Davina shares a story that explains her distaste for non-binary gender performances online. Although, she claims, people who are offline may know this about you, she questions why it is necessary online. For this participant, a change is confusing *because* Facebook compels users to perform "legal" or "real" identifications and *because* these performances are closely linked to official forms that rarely offer more than "female" and "male." It is clear that Facebook's gender affordances are both unused by, and confusing for, my participants.

I argue that these perceptions are in part because the more fluid gender options feel inconsistent with other affordances Facebook offers that the site links to "authenticity." For example, users, at sign up, still are provided only "female" and "male" from which to choose. They must also input their full birthday and their full, "real" name (as it appears on their birth certificate or driver's license). This type of rhetoric, paired with other options, prompted my informants to share that the increased gender options were probably just a pandering tactic.

From a programming perspective, the obvious heuristic is a model in which data are pulled for algorithmic ranking, personalized site character-istics, and third-party marketing first as "user gender is binary" or "user gender is not binary"; second as "user is female," "user is male," or "user is other"; and, as a third point, what is actually selected as the custom affiliation(s). In other words, users' digital gender identifications are still largely reliant on some binary estimation. On one hand, the new gender choice is appeasement; on the other, it serves as additional marketing data (Kellaway 2015).

Gender Just Matters More

In comparison to the added explicit space for gender performance, Facebook provides no such space that invites users to input their race/ethnicity identifi-cations. The absence of ethnicity is surprising considering how social media dramatize other identifications (as Facebook does—changes to gender tools were posted about and changed multiple times). The lack of race/ethnicity options speaks volumes about assumptions of the designers—there is a "cul-tural map of assumed whiteness" (Kolko 2000, 225). This "map" of whiteness is essential to understanding how designers' cultural expectations are baked into the spaces they create. At some point, someone decided that an explicit race/ethnicity category was not essential.

The omission of race/ethnicity space led my informants to share explana-tions that compared gender and race, almost as competing social issues. Three general themes emerged: (1) complexity, (2) post-racial culture, and (3) visual culture.

Complexity. Some of my informants spoke about Facebook's lack of racial identification space as rooted in the fact that race is more complex and compli-cated than gender. Therefore, some of them concluded that it makes sense that Facebook stays out of the conversation.

Dana, a 22-year-old white female: In our country, it wouldn't make sense. . . . There are so many different types of race and ethnicities.

Deb, an 18-year-old African American female: I know a lot of people who have a lot of race identity.

When thinking about Facebook race/ethnicity affordances, participants were quick to view them as too sticky to deal with, specifically when compared to gender. However, this way of thinking necessarily assumes that gender is not complex, even with the (new) knowledge that Facebook added 50+ gender affiliations. Users are at the risk of devaluing the struggles that many communities are experiencing in fighting for more than just "female" and "male." At the same time, these assumptions also put users at risk of thinking that, because an issue is complex, it means they should stay out. In sum, the identification choices provided perhaps devalue one fight while also pushing another to the sidelines. These perceptions are either carried from offline spaces and reified through Facebook or formed online by users who otherwise have not considered them.

Post-racial culture. As a second explanation, some participants discussed the possibility that race is purposely left out because Facebook is "color blind."

LJ, a 19-year-old white female: Umm, I like that there isn't one actually; I think that it's good that [Facebook's] color blind.

Ashley, a 20-year-old white female: I just don't think it [race] matters that much. It doesn't define you.

Ryan B., a 21-year-old white male: I'm just, I'm indifferent about it, I guess. I mean, it's something that I don't think, you know, represents the individual.

In line with other racial discourses present in the US, participants read Facebook's lack of explicit racial identification spaces as an indication that Facebook positions race as no longer a defining characteristic of people. Again, Facebook's omission promotes a very specific perception of racial affairs. Whether these are perceptions that started offline and were reified online, or they are perceptions that Facebook helped to cultivate, my participants were quick to assume that digital spaces are creators of new, equal environments.

In contrast, many argue that we learn to interact with another through physical features (e.g., Alcoff 2006; Nakamura and Chow-White 2012). As Nakamura (2002) argues, the internet is a space for *cybertyping*. Digital spaces harbor hegemonic ideals, and race becomes just as important online as it is offline (Martin, Trego, and Nakayama 2010; Tynes, Reynolds, and Greenfield 2004). Thus, it could be argued that Facebook works to support notions that the US is post-racial.

The point of this analysis is not to critique my informants for their opinions and interpretations of Facebook's goals. Rather, I intend to show one of many instances in which affordances online can have strong impacts on users. When considering how Facebook's interface is constantly constructing reality, highlighting specific traits and trends while squelching others,[6] we can conclude that many young users' views of the US are cultivated through digital affordances.

Visual culture. Ultimately, discussions with the focus groups led to considerations of our current, highly visual culture. Visible, corporeal identifiers, namely profile pictures, are seen as deeming explicit racial/ethnic affiliation unnecessary.

> *Stephanie, a 27-year-old white female*: I think it's the fact that you can post a picture of your race but you can't post a picture of your gender.
>
> *JM, a 20-year-old white female*: What I'm saying is you can see what they look like. And, if you want to know what their race is, you can ask them, sort of. But looking at them you really can't, like, sexuality doesn't have a color. I think people identify with that more, not to say that I agree with that; but I feel like that's how Facebook is saying it.

Throughout my focus groups, both when speaking directly about gender and race, and otherwise, being visible was integral. Users want to *see* those with whom they interact. There is a certain cultural anxiety that exists for many when they cannot decipher another's gender or race. This is perhaps rooted in the fact that first impressions help us to define people, and stereotypes aid in the process. As such, online functionalities exploit this "stereotypical shorthand" (Kolko 1999, 181). Stereotypes are easy because they quickly describe users and are easy to fit into databases. Designers' first choices are often the tools and functionalities that are derived from a very small selection of stereotypes (Kolko 1999).

In particular, Facebook promotes a very visible culture (Cirucci 2015), that begins with the profile photo. My informants shared many stories in which they were "creeped out" or annoyed when another Facebooker did not have a recognizable image of their face included in their profile photo. Users rely on these photos as a means to quickly summarize others' lives in a matter of minutes (Farquhar 2013). My participants labeled the profile picture as "prime real estate" and placed much value in its ability to not only add aesthetic value to a profile, but also provide identification validation. Many admitted to not interacting with faceless users even when *they know the person offline and know for certain that the profile is controlled by their offline friend.*

The adherence to, and expectation for, visible selves allows users to rely on visible tells for race/ethnicity. As Stephanie includes, there must be a gender selection because that can be "hidden" in a photo. But, you cannot "hide your race." In the case of my participants, the assumption that visible qualities make it easy to guess someone's race/ethnicity is driven by the site's omission of race and ethnicity categories, as well as their stark adherence to selves that are extremely visible.

In addition, it is intriguing to question why Facebook would leave out race/ethnicity seeing that it is a valuable marketing tool. Beyond the notion that Facebook employs powerful algorithms that likely abstract what a user's race/ethnicity is through stereotypical likes and browsing history, Stephanie may be correct. It is not impossible, or even improbable, that Facebook "guesses" race/

ethnicity based on profile, uploaded, and tagged photographs, among other data collected.

Gaver (1991) describes *hidden affordances* as functionalities that are afforded but with no information available about them. This is what Facebook essentially does—provides spaces that do not ask for race or ethnicity socially, while, at the same time, builds databases that still collect this information at the institutional level. In a sense, users are aiding in the negotiation of the affordances, without even realizing it. It is perhaps important to note here that what a "profile" or "user" looks like for a user is drastically different from what a "profile" or "user" looks like in Facebook's databases. Thus, it would be naïve to assume that just because Facebook makes a tool visible, that it matters greatly to the database, just as when an affordance is hidden it can still matter greatly. In sum, when an affordance is hidden by Facebook, it means that the full process and purpose are hidden, and the company does not want users to have much active say in the meaning negotiation.

Indeed, Facebook has implemented two processes in particular that are fairly hidden to the general user and that play a large role in how they categorize race and ethnicity—DeepFace and Multicultural Affinity Targeting. DeepFace is 3D modeling software that can verify faces with 97.35% accuracy, less than one percent away from human-level capabilities (Taigman et al. 2014). Thus, it is quite easy for Facebook to document skin color and other facial features that may align with racial and ethnic differences. There is no mention of DeepFace in Facebook's Help section when searching for and reading about what the site does to and with user images. In fact, when you search "deepface" in the Help Center, no matches are returned.

Multicultural Affinity Targeting helps advertisers, and Facebook, target people with "multicultural interests." They define multicultural affinity as "the quality of people who are interested in and likely to respond well to multicultural content" (Fussell 2016). As with Nakamura's (2002) early study of LambdaMOO where she shows that white is the "default" race and all others must be defined otherwise, white Facebookers do not have ethnic affinities—they are reserved for African American, Asian, and Latinx users (Thomas 2016). Thus, paired with DeepFace data, how users perform through the site allows Facebook to ascribe an "affinity." Then, advertisers can choose to *exclude* certain affinities from their marketing sample. This means that I could post an advertisement that excludes all users that have been deemed African American, Asian, and Latinx, with the hopes of my ad only being visible to white users. Facebook is careful to label these "affinities" and not ethnicities to avoid any lawsuits or bad press (Thomas 2016).

Above, I quote JM, who suddenly realized, toward the end of the focus group, that Facebook perhaps is sending an implicit message through the omission of explicit race/ethnicity spaces. This is representative of many of my participants' comments toward the end of their focus group. After much thought and conversation, some began to think that perhaps Facebook compels them to think about gender and race in specific ways. To be clear, just as Facebook cannot solve social divisions, it is not my goal to argue that the site is solely creating new social divisions. Clearly, the issues discussed herein are not new. However, in a space that has been so seamlessly folded into many lives, it is important to investigate what stereotypical norms are promoted.

Facebook's active decision to leave race/ethnicity off the user interface, while constantly updating and posting about gender affiliation, led my informants to view gender as a more important, but less messy, identification piece. Viewing US society as post-racial and guessing race/ethnicity through skin color and other physical "tells" are normalized through Facebook's functionalities—our identities are "necessarily shaped by platform design choices" (Lingel and Golub 2015, 547).

None of the Above

Finally, participants consistently shared experiences wherein they spoke to feeling like they had no choice—it is either be on Facebook and follow their rules or not use the space at all. It is true that space for transgressive performances is lacking. But, my participants were also highly unlikely to mention attempting to subvert the mainstream metanarratives offered by Facebook. Because of this, I asked them to consider why there is not more space allotted for agency. In particular, what about more "none of the above" spaces?

One common response across groups was to choose the privacy setting "only me." Users still fill the form as Facebook has designed, but "no one else" gets to see the choice. This is an interesting and relevant example of what Raynes-Goldie (2010) terms social privacy versus institutional privacy. At the social level—friends, networks, the public—any information marked as "only me" is hidden. However, at the institutional level—Facebook, third party marketers, database companies—these choices are still saved, used, filtered, and converted into profit. So, yes, Facebookers can hide their choices from their social networks, but these data are still blended into their online experiences, affecting ads displayed, news storied offered, and friends suggested.

Indeed, agency is often a tough issue when considering digital spaces because they are discrete, binary systems. Just as in a video game like *Super Mario Brothers*, say, where the gamer should have no expectation that Mario will be able to do something he has not been programmed to do (like move on the z-axis or be besties with Bowser), how social network sites are built, designed, and presented necessarily define what users can and cannot do. And, while structures do have the possibility of being transformed when users decide to take thoughtful action, the possibility of this kind of agency is often staged, made less visible, or quietly removed (Gillespie 2006).

As one example, when Facebook's About Page only offered binary gendered options, there was a bug that also, accidently, allowed users to alter a small line of code so that their gender would appear as neither female nor male. With some programming knowledge, this hack was easy to achieve. However, Facebook quickly patched the hole, and it is likely that that type of mistake, or bug, in the code will never occur again. Anonymity is also key to agency within digital systems (Magnet 2007), but, as discussed earlier, Facebook's drive for legal, official, and visible users is anything but anonymous.

What I really learned from these discussions is that there is a strong need for completion baked into Facebook's architecture. Some options are mandatory, like gender. Others, while perhaps not mandatory, are strongly suggested, and Facebook constantly reminds users that they have not completed or recently updated a section—perhaps a user has not updated their profile picture in sometime, and Facebook fears it no longer "accurately" represents the user's corporeal self. Thus, my informants were compelled to believe that Facebook is *official*. Although not a governmental or medical space, it is the *official social space*. They expect one another to be honest online, and the way to "enforce" this "authenticity" is to take the site's prompts as seriously as expected with filling out the US Census or a job application.

Options for gender and race were integral to these conversations. Because Facebook is official for my participants, and similar to other forms people fill out, the site "obviously" needs to collect this identifying information. This interpretation—that Facebook is official and calls for legal and corporeal data—is quite lucrative for the site. It ensures that Facebookers are *accurately*,[7] not necessarily authentically, broadcasting themselves.

> *JM, a 20-year-old white female:* I think when I filled out Facebook it was, like, so long that it was just kind of like, kind of like checking off a physical form, like, male, female, what are you interested in . . .
>
> *Alessia, a 20-year-old white female*: When you first start out with Facebook, it's an application process too . . .

The perception that Facebook is some official, patrolled space is in line with comments from its creator, Mark Zuckerberg:

> You have one identity. The days of you having a different image for your work friends or co-workers and for the other people you know are probably coming to an end pretty quickly. Having two identities for yourself is an example of a lack of integrity.
>
> (Kirkpatrick 2011, 199)

Immersed in the "official" Facebook space, users are compelled to believe that accuracy *is* authenticity. Thus, to perform "authentically," my participants were likely to input identifying information that is deemed important through Facebook's affordances. As such, to be "authentic" is to accurately include the gender that was assigned at birth and to not explicitly define racial or ethnic affiliations.

Facebook's promoted and morphed definition of authentic that has come to mean accurate, legal, and official is highlighted by a phenomenon seen through another Facebook platform—Instagram. Instagrammers create second, socially private, accounts known as Finstas (Williams 2016). In complete accordance with Facebook's definition of authentic, a Finsta—a *fake* Instagram account—is actually a more real representation of a user. The account is only shared with close friends and often includes embarrassing, unfiltered, mundane images. Thus, the "fakeness" of this type of account is not that it is inauthentic to the user, but that it is not in line with the definition of authentic that the architectures of spaces like Facebook, Instagram, and the like have cultivated (e.g., Papacharissi 2009).

Conclusion

It is perhaps becoming less and less a secret that Facebook strategically decides how identities will be shaped in an effort to construct more efficient data collection, algorithmic, and marketing models. The process of selecting which identification affiliations to request, and which to simply leave off the user interface, places value on specific identifications. As supported through my findings, Facebookers are led to adopt specific expectations and norms regarding the identification process and important cultural issues. As my focus group participants demonstrated, some believe that gender is a more important issue than race because Facebook explicitly asks users to define it. Others noted that race is a more complex and important fight than gender, and Facebook is right in "staying out." Thus, just as offline expectations follow us into online spaces, prejudices that we learn online journey with us into offline spaces—they are naturalized and reified through our constant, digital performances guided by the site's design.

Through the structural discourse analysis and focus groups, two main conclusions emerged. First, the negotiation of affordances, as defined by Gibson and updated for social network sites, is not, in fact *cannot* be, equal because the power roles at play at not equal. Facebook, as Gibson explained, creates a non-neutral space that makes more available what is beneficial to them. They control both the institutional data and the social interface. Facebook's employees decide how tools, functionalities, and buttons will be designed, how the data will be catalogued and saved, and what will happen to them over time. They decide which data are important and which are "throwaway," included at the social level for user appeasement. Thus, while users are certainly allowed to do as much as they can within the site, they play only a small role in what the affordances are. This is made especially clear by the way in which my participants view the site as an "official" space or a social utility. Just as they would not want to lie on a form, they do not want to cheat the Facebook system.

The second conclusion situates my sample within the generally heteronormative Facebooker type—when users affiliate with more privileged and socially accepted identifications (whether by choice or through social pressure), they are not inspired to tinker with the site or resist the norms being cultivated. This would at first seem counter-intuitive—those with more social *power* should have more power in the negotiation of affordances. However, those with marginalized identities are *used* to fighting against how they are shoved into categories in spaces *exactly* like job applications or medical forms.

Marginalized individuals, especially when considering gender and race/ethnicity affiliations, are not concerned with being "accurate" because many official forms do not even provide them with the correct spaces and options to be "accurate." For these groups, being authentic still means defying what forms present as identifications. Therefore, it is clear that when a part of privileged identity groups, as were most of my participants, users are not likely to even think about how they could, or should, subvert the Facebook architecture from the inside. Facebook's functionalities and policies reflect particular assumptions of identity that privilege some users over others. But, it is those who are marginalized that attempt to find workarounds (Lingel and Golub 2015).

The structural rules implemented by Facebook, although not always designed as visible affordances at the user-level, both set parameters for what is possible (Hutchby 2001b) while also compelling users to act in particular ways and, in turn, implicitly support and adhere to heteronormative identity expectations. It is not that each user is *determined* by Facebook's structure, but that, through a "regulated process of repetition that both conceals itself and enforces its rules precisely through the production of substantializing effects," users are molded (Butler 2006, 198). Agency, then, can only be located in some break of that repetition. This subversion is difficult, however, because unlike dressing in drag offline, if a cisfemale user decides to upload a profile photo wherein she is dressed in a stereotypically masculine way, Facebook will have enough other data points to continue to view, and market to, her as "female." In addition, until mainstream media break down the complex negotiation of digital affordances, most users will remain comfortable in, or at the very least unaware of, Facebook's promoted culture.

Notes

1. Although the main structural discourse analysis took place in January 2014, small pieces were added and changed throughout the course of this project as the interface changed. This study represents but a snapshot of time because Facebook's interface is constantly being updated.
2. http://zuckerbergfiles.org/
3. February 13, 2014.
4. At the time of publication, Facebook allows users to type anything in the "gender" box, after they have selected "custom" as their main gender, instead of "female" or "male."
5. Each informant was asked to choose a pseudonym and was provided a blank space to provide identifying information including, but not limited to: age, racial affiliation, ethnicity, gender, and socio-economic class.
6. It was reported that Facebook took down an iconic photo, "Napalm Girl," from the Vietnam War (Wong, 2016).
7. In this context, "accurately" is defined generally as performing some legal and corporeal self.

Works Cited

Alcoff, L. M. 2006. *Visible Identities: Race, Gender and the Self.* Oxford: Oxford University Press.

Bowker, G., and S. L. Star. 1999. *Sorting Things Out. Classification and Its Consequences.* Cambridge, MA: MIT Press.

boyd, d. 2014. *It's Complicated: The Social Lives of Networked Teens.* New Haven, CT: Yale University Press.

Butler, J. 2006. *Gender Trouble. Feminism and the Subversion of Identity.* New York: Routledge.

Cirucci, A. M. 2013. "First Person Paparazzi: Why Social Media Should Be Studied More Like Video Games." *Telematics and Informatics* 30 (1): 47–59.

———. 2015. "Facebook's Affordances, Visible Culture, and Anti-Anonymity." *Proceedings of the 2015 International Conference on Social Media & Society.* doi:10.1145/2789187.2789202.

Duguay, S. 2015. "Is Being #Instagay Different From an #lgbttakeover? A Cross-Platform Investigation of Sexual and Gender Identity Performances." In *SM&S: Social Media and Society 2015 International Conference.* Toronto: Ted Rogers School of Management, Ryerson University, July 27–29.

———. 2016. "He Has a Way Gayer Facebook Than I Do: Investigating Sexual Identity Disclosure and Context Collapse on a Social Networking Site." *New Media & Society* 18 (6): 891–907.

Fairclough, N. 1995. *Media Discourse.* London: Arnold.

Farquhar, L. 2013. "Performing and Interpreting Identity Through Facebook Imagery." *Convergence* 19 (4): 446–71.

Frey, J. H., and A. Fontana. 1993. "The Group Interview in Social Research." In *Successful Focus Groups: Advancing the State of the Art,* edited by D. K. Mogan, 175–87. Newbury Park: Sage Publications.

Fussell, S. 2016. "Facebook Might Be Assigning You an 'Ethnic Affinity' You Can't Change." *Fusion,* October. http://fusion.net/facebook-might-be-assigning-you-an-ethnic- affinity-you-1793863259.

Galloway, A. R. 2013. *The Interface Effect.* Malden, MA: Polity Press.

Gaver, W. W. 1991. "Technology Affordances." In *Proceedings of the SIGCHI Conference on Human Factors in Computing Systems,* edited by S. P. Robertson, G. M. Olson, and J. S. Olson, 79–84. New York: ACM, April.

Gibson, J. J. 1979. *The Ecological Approach to Visual Perception.* Boston, MA: Houghton Mifflin.

Giddens, A. 1984. *The Constitution of Society: Outline of the Theory of Structuration.* Berkeley, CA: University of California Press.

Gillespie, T. 2006. "Designed to 'Effectively Frustrate': Copyright, Technology and the Agency of Users." *New Media & Society* 8 (4): 651–69.

Gleit, N. 2008. "He/She/They: Grammar and Facebook." *Facebook,* June. www.facebook.com/notes/facebook/heshethey-grammar- andfacebook/21089187130.

Hutchby, I. 2001a. *Conversation and Technology: From the Telephone to the Internet.* Malden, MA: Blackwell.

———. 2001b. "Technologies, Texts and Affordances." *Sociology: The Journal of the British Sociological Association* 35 (2): 441.

Kellaway, M. 2015. "Facebook Now Allows Users to Define Custom Gender." *Advocate,* February. www.advocate.com/politics/transgender/2015/02/27/facebook-now-allows- users-define-custom-gender.

Kirkpatrick, D. 2011. *The Facebook Effect: The Inside Story of the Company That Is Connecting the World.* New York: Simon & Schuster.

Kolko, B. E. 1999. "Representing Bodies in Virtual Space: The Rhetoric of Avatar Design." *The Information Society* 15 (3): 177–86.

———. 2000. "Erasing @race: Going White in the (Inter)face." In *Race in Cyberspace*, edited by B. Kolko, L. Nakamura, and G. B. Rodman, 213–32. New York: Routledge.

Latour, B. 2005. *Reassembling the Social: An Introduction to Actor-Network Theory.* Cambridge, MA: Harvard University Press.

Light, B. 2007. "Introducing Masculinity Studies to Information Systems Research: The Case of Gaydar." *European Journal of Information Systems* 16 (5): 658–65.

Light, B., and K. McGrath. 2010. "Ethics and Social Networking Sites: A Disclosive Analysis of Facebook." *Information Technology & People* 23 (4): 290–311.

Lingel, J., and A. Golub. 2015. "In Face on Facebook: Brooklyn's Drag Community and Sociotechnical Practices of Online Communication." *Journal of Computer-Mediated Communication* 20 (5): 536–53.

Magnet, S. 2007. "Feminist Sexualities, Race and the Internet: An Investigation of Suicidegirls.com." *New Media & Society* 9 (4): 577–602.

Martin, J. N., A. B. Trego, and T. K. Nakayama. 2010. "College Students' Racial Attitudes and Friendship Diversity." *The Howard Journal of Communication* 21: 97–118. doi:10.1080/10646171003727367.

McNicol, A. 2013. "None of Your Business? Analyzing the Legitimacy and Effects of Gendering Social Spaces Through System Design." In *Unlike Us Reader: Social Media Monopolies and Their Alternatives. INC Reader #8*, 200–19. Amsterdam: Institute of Network Cultures. www.exhipigeonist.net/files/Unlike%20Us%20 Reader%20- %20Social%20Media%20Monopolies%20And%20Their%20Alternatives.pdf#page=202.

McVeigh-Schultz, J., and N. K. Baym. 2015. "Thinking of You: Vernacular Affordance in the Context of the Microsocial Relationship App, Couple." *Social Media+ Society* 1 (2): 2056305115604649.

Nagy, P., and G. Neff. 2015. "Imagined Affordance: Reconstructing a Keyword for Communication Theory." *Social Media+ Society* 1 (2): 2056305115603385.

Nakamura, L. 2002. *Cybertypes: Race, Ethnicity, and Identity on the Internet.* New York: Routledge.

Nakamura, L., and P. A. Chow-White. 2012. "Introduction—Race and Digital Technology: Code, the Color Line, and the Information Society." In *Race After the Internet*, edited by L. Nakamura and P. A. Chow-White, 1–18. New York: Routledge.

Norman, D. A. 1988. *The Psychology of Everyday Things.* New York: Basic Books.

Papacharissi, Z. 2009. "The Virtual Geographies of Social Networks. A Comparative Analysis of Facebook, LinkedIn, and a Small World." *New Media & Society* 11 (1–2): 199–220. doi:10.1177/1461444808099577.

Raynes-Goldie, K. 2010. "Aliases, Creeping, and Wall Cleaning: Understanding Privacy in the Age of Facebook." *First Monday* 15 (1). http://firstmonday.org/ojs/index.php/fm/article/viewArticle/2775/2432.

Rheingold, H. 1996. "A Slice of My Life in My Virtual Community." *High Noon on the Electronic Frontier: Conceptual Issues in Cyberspace*, 413–36.

Taigman, Y., M. Yang, M. Ranzato, and L. Wolf. 2014. "DeepFace: Closing the Gap to Human-Level Performance in Face Verification." In *Proceedings of the 2014 IEEE Conference on Computer Vision and Pattern Recognition* (CVPR), June. www.facebook.com/publications/546316888800776/.

Thomas, D. 2016. "Facebook Doesn't Know You're White." *Vice News*, November. https:// news.vice.com/story/facebook-tracks-your-ethnic-affinity-unless-youre-white.

Turkle, S. 1995. *Life on the Screen. Identity in the Age of the Internet*. New York: Simon & Schuster.

Tynes, B., L. Reynolds, and P. M. Greenfield. 2004. "Adolescence, Race, and Ethnicity on the Internet: A Comparison of Discourse in Monitored vs. Unmonitored Chat Rooms." *Journal of Applied Developmental Psychology* 25 (6): 667–84. doi:10.1016/j.appdev/2004.09.003.

Van Dijck, J. 2013. *The Culture of Connectivity: A Critical History of Social Media*. Oxford: Oxford University Press.

Williams, L. 2016. "Rinstagram or Finstagram? The Curious Duality of the Modern Instagram User." *The Guardian*, September. www.theguardian.com/technology/2016/ sep/26/rinstagram-finstagram-instagram- accounts.

Winner, L. 1980. "Do Artifacts Have Politics?" *Daedalus* 109 (1): 121–36.

Wong, J. C. 2016. "Mark Zuckerberg Accused of Abusing Power After Facebook Deletes 'Napalm Girl' Post." *The Guardian*, February. www.theguardian.com/technology/2016/ sep/08/facebook-mark-zuckerberg- napalm-girl-photo-vietnam-war.

Zhao, S., S. Grasmuck, and J. Martin. 2008. "Identity Construction on Facebook: Digital Empowerment in Anchored Relationships." *Computers in Human Behavior* 24 (5): 1816–36. doi:10.1016/j.chb.2008.02.012.

Appendix C
Industry Report Example

Usability Testing Summary and Redesign Recommendations

Petz.com Registration Page Redesign

Background

Petz.com is a one-stop online shop for pet owners to buy supplies—food, toys, medication, etc.—for their pets. As part of ongoing Petz.com website improvement initiative (see *Petz.com Website Improvements List*), the "customer registration page" has been identified as a key part of the website to redesign for usability. The purpose of this research was to understand how new users to the website navigate through the registration page to sign up to be regular customers of Petz.com.

Stakeholders

Sam Smith—UX Researcher
Dolly Rhea—UX Designer
Silvio Buresco—Software Developer
Poppy Lupus—Product Manager
Eddie Wharton—CEO

Methodology

Usability testing

Participants

Ten pet owners who were interested in signing up with Petz.com to buy pet supplies regularly. Participants were recruited through popover surveys on the website shown only to new visitors on their first visit (before registration).

Key Findings

Key findings from usability tests are summarized:

- Participants had a hard time locating the "register now" button.
- During registration, 2 out of 10 participants mistyped their email address by not using the @ symbol.
- Four participants were not comfortable with sharing their date of birth upon registration (they did not understand why this personal information was needed).

Known Limitations

The following is a known limitation to the data collected:

- Small sample size

Recommendations

Key recommendations are summarized:

- Increase the size of the "register now" button, and make it a darker color, so it stands out more and is easier to find on the home page.
- Show an error message alerting users of the mistake, that they did not input the correct email address format (missing an @).
- Make the "date of birth" data collection box optional, and provide a tooltip to explain why this data is being collected (this feature is used to send loyal customers "happy birthday" messages along with a discount code to use on their birthday, but it is not mandatory for registering as a regular shopper on the website).

Reference

Petz.com Website Improvements List, gathered from complaints from previous users

Appendix

Instructions and script for usability testing

Index